About the Authors

Thomas Eugene Svarney and **Patricia Barnes-Svarney** have backgrounds in the physical sciences. Svarney is currently a nonfiction writer and researcher; Barnes-Svarney is a nonfiction and science fiction/mystery writer. Research for their books has led them to all corners of the world—from Meteor Crater, Arizona, and dinosaur digs in Colorado to naturalists studies in Europe, and even weather research at the South Pole.

Barnes-Svarney has written several books, including the award-winning bestseller, *New York Public Library Science Desk Reference,* and Book-of-the-Month pick *Asteroid: Earth Destroyer or New Frontier?* She has more than 350 article credits in such magazines as *Popular Science, Air & Space,* and *Weatherwise*. Together, Svarney and Barnes-Svarney have co-authored many adult science books, including four for Visible Ink's "Handy Answer Book" series (Dinosaurs, Oceans, Geology, and Math), as well as *A Paranoid's Ultimate Survival Guide* and *Skies of Fury: Weather Weirdness around the World.*

Also from Visible Ink Press

The Handy Anatomy Answer Book
by James Bobick and Naomi Balaban
ISBN: 978-1-57859-190-9

The Handy Answer Book for Kids (and Parents)
2nd edition
by Gina Misiroglu
ISBN: 978-1-57859-219-7

The Handy Astronomy Answer Book
by Charles Liu
ISBN: 978-1-57859-193-0

The Handy Biology Answer Book
by James Bobick, Naomi Balaban, Sandra Bobick, and Laurel Roberts
ISBN: 978-1-57859-150-3

The Handy Geography Answer Book
2nd edition
by Paul A. Tucci and Matthew T. Rosenberg
ISBN: 978-1-57859-215-9

The Handy Geology Answer Book
by Patricia Barnes-Svarney and Thomas E Svarney
ISBN: 978-1-57859-156-5

The Handy History Answer Book
2nd edition
by Rebecca Nelson Ferguson
ISBN: 978-1-57859-170-1

The Handy Math Answer Book
by Patricia Barnes-Svarney and Thomas E Svarney
ISBN: 978-1-57859-171-8

The Handy Ocean Answer Book
by Patricia Barnes-Svarney and Thomas E Svarney
ISBN: 978-1-57859-063-6

The Handy Philosophy Answer Book
by Naomi Zack
ISBN: 978-1-57859-226-5

The Handy Physics Answer Book
by P. Erik Gundersen
ISBN: 978-1-57859-058-2

The Handy Politics Answer Book
by Gina Misiroglu
ISBN: 978-1-57859-139-8

The Handy Religion Answer Book
by John Renard
ISBN: 978-1-57859-125-1

The Handy Science Answer Book™
Centennial Edition
by The Science and Technology Department Carnegie Library of Pittsburgh
ISBN: 978-1-57859-140-4

The Handy Sports Answer Book
by Kevin Hillstrom, Laurie Hillstrom, and Roger Matuz
ISBN: 978-1-57859-075-9

The Handy Supreme Court Answer Book
by David L Hudson, Jr.
ISBN: 978-1-57859-196-1

The Handy Weather Answer Book
by Kevin Hile
ISBN: 978-1-57859-221-0

THE HANDY DINOSAUR ANSWER BOOK

THE HANDY DINOSAUR ANSWER BOOK

SECOND EDITION

Patricia Barnes-Svarney
and Thomas E. Svarney

Detroit

THE HANDY DINOSAUR ANSWER BOOK

Visible Ink Press®
43311 Joy Rd., #414
Canton, MI 48187-2075

Visible Ink Press is a registered trademark of Visible Ink Press LLC.

Most Visible Ink Press books are available at special quantity discounts when purchased in bulk by corporations, organizations, or groups. Customized printings, special imprints, messages, and excerpts can be produced to meet your needs. For more information, contact Special Markets Director, Visible Ink Press, www.visibleink.com, or 734-667-3211.

Managing Editor: Kevin S. Hile
Art Director: Mary Claire Krzewinski
Typesetting: Marco Di Vita
Proofreader: Sharon R. Malinowski

ISBN 978-1-57859-218-0

Front cover images: iStock. Back cover images: Big Stock Photo.

Library of Congress Cataloging-in-Publication Data

Barnes-Svarney, Patricia L.
The handy dinosaur answer book / Patricia Barnes-Svarney and Thomas E. Svarney. — 2nd ed.
p. cm.
Includes bibliographical references and index.
ISBN 978-1-57859-218-0
1. Dinosaurs—Miscellanea. I. Svarney, Thomas E. II. Title.
QE861.4.S83 2010
567.9—dc22

2009032573

Printed in the Thailand

10 9 8 7 6 5 4 3 2 1

Contents

Introduction

Ten years is a long time in science; there are so many discoveries and new technologies that change and add to the overall scientific field. It's no different in paleontology—especially in the study of dinosaurs.

So now, for you, the dinosaur enthusiast, we present the new, colorful, and very updated second edition of *The Handy Dinosaur Answer Book*. One of the best parts of revising and updating a book is going through all the feedback we've heard/read over the years. And there's no doubt about it: this book seems to be a favorite of dinosaur lovers from 6 to 96!

One of the reasons for this fascination with the great creatures has not changed in 10 years: How in the world did something as big as a house—and sometimes ferocious—roam Earth so many millions of years ago? Admittedly, the land has changed over the past 65 million years since the dinosaurs died out, but just to think that an *Allosaurus* or *Gigantosaurus* may have once stomped in your own backyard never fails to send chills up our collective human spines.

There are other reasons for dinosaur enthusiasm, especially with so many global discoveries made in the past decade: the discovery of fossilized dinosaur organs; whole or parts of dinosaur eggs and nests; imprints of dinosaur blood vessels; feather imprints galore; attempts to extract DNA from dinosaur remains; and large caches of fossils in rock formations previously thought to be devoid of dinosaur bones. There are also more and more large-boned dinosaur fossils that pop up every year, keeping alive the "size competition" between the big carnivores like the *Tyrannosaurus rex* and sauropods like the *Brachiosaurus*. How could anyone who's a dinosaur enthusiast not be excited about such amazing discoveries?

This newly revised *The Handy Dinosaur Answer Book* answers over 600 of your questions about these extinct creatures, such as: From what animals did the dinosaurs evolve? Who found the first dinosaur remains? What other plants and animals lived during the time of the dinosaurs? What dinosaur remains show a heart? What are the largest—and smallest—dinosaur fossils known at this time? How did scientists extract DNA from a *Tyrannosaurus rex* bone? Are birds really dinosaurs? Did the first dinosaurs have feathers? Why are Argentina and China important to recent dinosaur studies? And the question and answer list goes on and on.

This book attempts to answer these questions and more, taking you through the Triassic, Jurassic, and Cretaceous periods of geologic time when the dinosaurs roamed the earth. It tells you where scientists stand in terms of dinosaur discoveries, the newer species of dinosaurs, new possible reasons for the extinction of the animals, and especially why, as more dinosaur fossils are uncovered, the ideas about the evolution of dinosaurs will continue to change.

We invite you to enjoy reading about the most amazing creatures that ever walked, ran, hopped, loped, and stomped on the Earth. As the saying goes: "Dinosaurs Once Ruled!" If the current fascination with dinosaurs is any indication, they still do.

—Patricia Barnes-Svarney and Thomas E. Svarney

Acknowledgments

This book would not be possible without the hard work and dedication of the people at Visible Ink Press. Our thanks especially go to our editor, Kevin Hile, for his wonderful editing and insightful questions, and for making this dinosaur book so user-friendly! We also want to thank our agent, Agnes Birnbaum for all her help, patience, and friendship over the years.

—Patricia Barnes-Svarney and Thomas E. Svarney

FORMING FOSSILS

IN THE BEGINNING

How **old** is **Earth**?

Earth is currently believed to be about 4.54 billion years old, but that number came after centuries of debate. In 1779, French naturalist Comte de Georges Louis Leclerc Buffon (1707–1788) caused a stir when he announced 75,000 years had gone by since Creation, the first time anyone had suggested that the planet was older than the biblical reference of 6,000 years. By 1830, Scottish geologist Charles Lyell (1797–1875) proposed that Earth must be several hundred million years old based on erosion rates; in 1844, British physicist William Thomson, later first baron of Largs (Lord) Kelvin, (1824–1907), determined that Earth was 100 million years old, based on his studies of the planet's temperature. In 1907, American chemist and physicist Bertram Boltwood (1870–1927) used a radioactive dating technique to determine that a specific mineral was 4.1 billion years old (although later on, with a better knowledge of radioactivity, the mineral was found to be only 265 million years old). Using different adaptations of Boltwood's methods on terrestrial, lunar, and meteorite (space rock that falls to the surface of Earth) material, scientists now estimate that Earth is between 4.54 and 4.567 billion years old.

How old is the **oldest rock and mineral** found on Earth?

The oldest rock discovered on Earth, the Acasta gneisses found in the tundra in northwestern Canada near the Great Slave Lake, is about 4.03 billion years old. The oldest minerals yet found are 4.404 billion years old and were found in Western Australia. The minerals—zircon crystals—eroded from their original rock, and then were deposited in younger rock.

Gases released by erupting volcanoes, such as carbon dioxide, nitrogen, and water vapor, did a great deal during Earth's early history to make the atmosphere life-sustaining for plants and animals (iStock).

What **caused** the early Earth's **water** and **atmosphere** to **form**?

No one really knows how the oceans filled with water. One theory is that volcanoes released enough water vapor to allow the oceans' waters to condense. Another theory states that comets bombarded Earth just after the formation of the solar system, bringing enough water to eventually fill the oceans.

The origin of Earth's atmosphere is also debated, but not as intensely. In this case, it is more likely that some of the atmosphere originated from gases that were part of the solar nebula, gases brought by comets, and those produced from volcanic activity. Earth probably would have had a thicker atmosphere, too, but the young, active Sun's heat boiled away the lighter materials—elements that are still found today around the gas giant planets Jupiter, Saturn, Uranus, and Neptune.

What **gases** began to **accumulate** after Earth's crust finally solidified?

As Earth's crust solidified, gases began pouring out of fissures and volcanoes, accumulating in the forming atmosphere. These same gases still emanate from modern volcanoes, and include carbon dioxide (CO_2), water vapor (H_2O), carbon monoxide (CO), nitrogen (N_2), and hydrogen chloride (HCl).

As these gases interacted in the atmosphere, they combined to form hydrogen cyanide (HCN), methane (CH_4), ammonia (NH_4), and many other compounds. This atmosphere would be lethal to most present day life-forms. Fortunately for life on

Lilies are grown in a greenhouse in Almere, the Netherlands. Just as this structure allows tropical plants to grow in a cold climate, the natural greenhouse effect created by Earth's atmosphere warms our planet (iStock).

Earth, over the next two to three billion years the atmosphere continued to change until it reached close to its present composition.

How did **oxygen** form on **early Earth**?

The early atmosphere was composed mainly of water vapor, carbon dioxide and monoxide, nitrogen, hydrogen, and other gases released by volcanoes. By about 4.3 billion years ago, the atmosphere contained no oxygen and about 54 percent carbon dioxide. About 2.2 billion years ago, plants in the oceans began to produce oxygen by photosynthesis, which involved taking in carbon dioxide. By two billion years ago, there was one percent oxygen in the atmosphere, and plants and carbonate rocks caused carbon dioxide levels to decline to only four percent. By about 600 million years ago, atmospheric oxygen continued to increase as volcanoes and climate changes buried a great deal of plant material—plants that would have absorbed oxygen from the atmosphere if they had decomposed in the open. Today, our planet's atmosphere levels measure 21 percent oxygen, 78 percent nitrogen, and only 0.036 percent carbon dioxide.

What is the **greenhouse effect**?

The greenhouse effect, as its name implies, describes a warming phenomenon. In a greenhouse structure, closed glass windows cause heat to become trapped inside. The greenhouse effect functions in a similar manner, but on a planetary scale. In general, it occurs when the planet's atmosphere allows heat from the Sun to enter but refuses to let it leave.

Without this greenhouse effect on Earth, life as we know it would not exist. On our planet, solar radiation passes through the atmosphere and strikes the surface.

Why is global warming important to humans?

The scientific consensus is that global average temperatures are rising, a phenomenon often referred to as global warming. Many scientists believe human activity has greatly contributed to the buildup of greenhouse gases in Earth's atmosphere in the past century or so, and hence Earth's gradual warming—around 1 degree Fahrenheit (0.5° Celsius). One recent study by an international panel of scientists predicted that the global average temperature could increase between 2.5 and 10.4° Fahrenheit (1.4 and 5.8° Celsius) by the year 2100 and that sea levels could rise by up to 2 feet (just over a half meter).

What is the biggest culprit? Although there are other gases, such as methane and chlorofluorocarbons, that increase global warming, most experts point to carbon dioxide as the worst pollutant in this case. This gas is released into the atmosphere mainly through burning of fossil fuels, such as coal, gasoline, and diesel. The gas also forms from the destruction of natural vegetation, such as the burning of forests to turn into grazing meadows for livestock. In this case, the carbon dioxide releases in two ways. First, the destruction of plant life through human actions causes less carbon dioxide to be absorbed out of the atmosphere; and secondly, rotting vegetation in clear-cut forests releases carbon dioxide.

As it is reflected back toward space, some solar radiation is trapped by atmospheric gases such as carbon dioxide, methane, chlorofluorocarbons, and water vapor, resulting in the gradual increase of Earth's temperatures. The rest of the radiation escapes back into space. Without this heat, life as we know it would be impossible, Earth would be about 100 degrees cooler, and the oceans would freeze.

What is **ozone** and how did it **benefit** the **early Earth**?

Ozone (O_3)—compared to the oxygen (O_2) we breathe—usually refers to a blanket of gas found between 9 and 25 miles (15 and 40 kilometers) in the layer of Earth's atmosphere called the stratosphere. The so-called "ozone layer" is produced by the interaction of the Sun's radiation with certain air molecules. The blue-tinged ozone gas is also found in the lower atmosphere. While beneficial in the stratosphere, ozone forms photochemical smog at ground level. This smog is a secondary pollutant produced by the photochemical reactions of certain air pollutants, usually from industrial activities and cars.

The stratosphere's ozone layer is important to all life on the planet because it protects organisms from the Sun's damaging ultraviolet radiation. Scientists believe that about two billion years ago, oxygen was being produced by shallow water marine plants. This sudden—geologically speaking—outpouring of oxygen helped to build up the ozone layer. As the oxygen levels increased, ocean animals began to evolve.

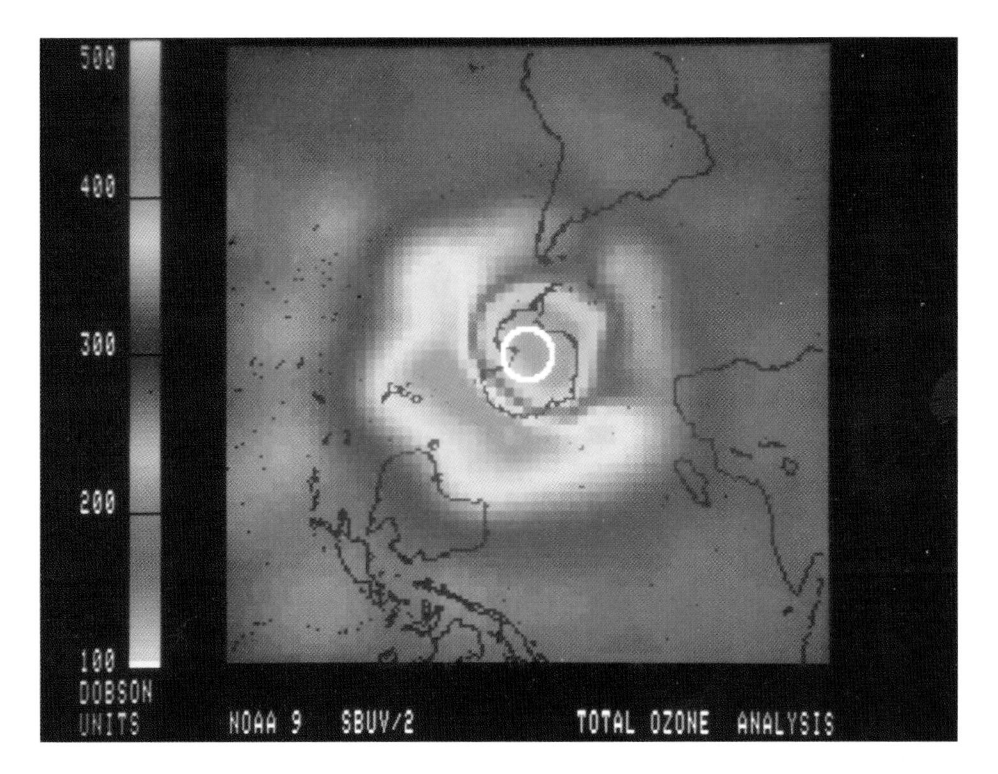

The formation of the ozone layer in the upper atmosphere early in Earth's history created a radiation boundary that protects life on the planet. Today, scientists are concerned about the hole in the ozone that has appeared over the South Pole, as seen in this 1987 satellite image (National Oceanographic and Atmospheric Administration).

Once the protective ozone layer was in place in the atmosphere, it allowed the marine plants and animals to spread onto land, safe from the Sun's radiation.

BEGINNINGS OF LIFE

When did **life first begin** on Earth?

No one knows the precise time that life began on Earth. One reason is that early life consisted of single-celled organisms. Because the soft parts of an organism are the first to decay and disappear after death, it is almost impossible to find the remains. In addition, because the organisms were so small, they are now difficult to detect in ancient rocks. Some modern viruses are only about 18 nanometers (18 billionths of a meter) across and modern bacteria typically measure 1,000 nanometers across, which is much larger than the early organisms.

In addition, because scientists have found so little fossil evidence, it is difficult to know all the true shapes of the earliest life. Scientists believe that early life was composed of primitive single-cells and started in the oceans. The reason is simple:

Could life have arrived from outer space?

There is another theory of how the precursors of life were brought to Earth—known as panspermia. Scientists theorize that comets and asteroids bombarded the early Earth, bringing complex organic materials, many of which survived the fall to our planet.

Scientists know there are such organic materials in space. In the late 1960s, radio astronomers discovered organic molecules in dark nebulae. Since that time, other sources have been found, including organic molecules existing in space bodies such as asteroids, comets, and meteorites. In 1969, analysis of a meteorite showed at least 74 amino acids within the chunk of rock. Scientists began to speculate that the organic molecules could have traveled to Earth via meteorites, cometary dust, or, during the early years of Earth, by way of comets and asteroids.

Although many scientists argue that the heat from the impact of a giant asteroid or comet would destroy any organic passengers, many other scientists disagree. They propose that only the outer layers of a large body would be affected, or that the fine, unheated dust of comets could have brought the necessary amino acids to Earth. If this theory is true, we are apparently all—from dinosaurs to humans—made of "star stuff."

life needed a filter to protect it from the incoming ultraviolet energy from the Sun—and the ocean waters gave life that protection.

Despite such gaps in knowledge, scientists estimate that the first life began about four billion years ago. These organisms did not survive on oxygen, but carbon dioxide.

What were the **conditions** on the early Earth that scientists believe may have **led to life**?

Two major theories explain how life could have grown on early Earth. The first theory states that life grew from a primordial "soup," a thick stew of biomolecules and water. Chemical reactions were then triggered by the Sun's ultraviolet rays, lightning, or perhaps even the shockwaves from violent meteor strikes that were more common at the time. These reactions produced various carbon compounds, including amino acids, which make up the proteins found in all living organisms. This theory was postulated after a famous experiment performed at the University of Chicago in 1954 by then-graduate student Stanley Miller (1930–2007), and his advisor, chemist Harold Urey (1893–1981). They showed that the amino acids could be formed from chemicals thought to exist in the early Earth atmosphere when they were combined with water and zapped by lightning.

One of the earliest forms of life to appear on Earth was cyanobacteria (inset), which have left behind unusual fossil rocks called stromatolites. (iStock).

The second theory of life centers around a discovery made within the last half century: hydrothermal vents, which are cracks caused by volcanic magma seeping through the deep ocean floor. There were probably many more hydrothermal vents during the early history of Earth, as the crust was newer, and thus thinner, than today's cooled, thicker crust. The organisms around these vents did not need to rely on photosynthesis for energy. Scientists know that today's volcanic vent organisms live off the bacteria around the vents, which in turn extract energy from the hot, hydrogen sulfide-rich water found around the sunless cracks in the ocean floor. Early organisms could have survived in much the same way.

In actuality, the conditions described by both theories could have existed simultaneously to produce the planet's early life.

What are the **oldest-known fossils** found in rock on Earth?

The oldest-known fossils in rock have been found in Australia. One set of fossils found in Western Australia is dated between 3.45 and 3.55 billion years old. They show evidence of layered mounds of limestone sediment called stromatolites, which were formed by primitive microorganisms similar to blue-green algae called cyanobacteria. Scientists know that stromatolites exist today. The fossils look amazingly like the stromatolites from the shallow waters off the coast of modern Australia.

There are other contenders for the oldest-known fossils. Tiny, simple cells have also been found in ancient cherts (crystalline-rich sedimentary rocks) from Australia, and there are similar ones in Africa. These cells are preserved by the silica from the chert, and appear to show a cell wall of some kind.

When did the **basic form** of life **develop** on Earth?

It is thought that as far back as about 3.8 billion years ago, a basic form of life was present on Earth. This life took the form of tiny cells, which were surrounded by membranes to isolate and protect their interiors from the surrounding environment. The cells had a basic genetic system similar to those in modern cells, and this allowed the cells to self-replicate. We classify these earliest life-forms as prokaryotes, which includes such organisms as bacteria and cyanobacteria.

When did **larger cells develop**?

Larger cells, classified as eukaryotes, began to develop approximately 1.5 to 1.9 billion years ago, according to the known fossil record. Before this time, rock layers contained only tiny prokaryotes, such as bacteria and blue-green algae.

When did the **first multicellular** forms of life develop?

Based on the known fossil record, the first true primitive forms of multicellular life apparently developed around 650 million years ago, although some scientists classify a certain 1.2 billion-year-old red algae as a "taxonomically resolved" multicellular organism. (Humans are considered multicellular organisms, complete with 100 trillion cells that make up our bodies.)

One of the first such organisms is thought to have been a primitive form of sponge. The first fossil records of burrows are also found around the same time. These multicellular organisms are called Ediacaran fauna or assemblages (they are named after the Ediacaran hills in Southern Australia). Most have large surface areas, perhaps in response to their need to absorb oxygen, as there were very small concentrations of this gas present in the atmosphere at that time. They appear to have lived in shallow marine environments.

Did **life develop** more than just **one time**?

Many scientists believe that life may have started over and over on Earth. They speculate that once life began—either around ocean vents and/or in the shallow seas—comets and asteroids would strike the planet, killing off all the beginning stages of life. This may have happened many times over millions of years, until life became stable enough to sustain and diversify itself.

When did the **first true plants** appear on **land**?

Fossils called *Cooksonia,* found in Ireland, were probably the first true macroscopic plants to colonize land about 425 million years ago. Other plants also appeared not long after, including flowerless mosses, horsetails, and ferns. They reproduced

This yellow tube sponge, found near the Cayman Islands, descends from sponges that were among the first multicellular life on the planet (iStock).

by throwing out spores or minute organisms that carried the genetic blueprint for the plant. The ferns eventually developed seeds, but this did not happen until about 345 million years ago. Vascular plants—those with roots, stems, and leaves—evolved about 408 million years ago.

When did the **first soft-bodied animals** appear in the oceans?

Fossils reveal that the first soft-bodied animals appeared about 600 million years ago in the oceans. They included a form of jellyfish, as well as segmented worms.

What is the **oldest-known life form** that existed on **land**?

So far, there is no definitive agreement about the oldest-known land life to have emerged on Earth, but there have been some intriguing discoveries. For example, in 1994 scientists in Arizona discovered fossilized tubular microorganisms dating back 1.2 billion years. In 2000, another team of scientists at NASA's Astrobiology Institute uncovered an even older possibility: fossilized remnants of microbial mats (composed primarily of cyanobacteria) that developed on land between 2.7 billion and 2.6 billion years ago in the eastern Transvaal district of South Africa. Around 2002, yet another scientist uncovered what he thought may have been the earliest life on land in the form of a biocrust—a thin film of bacteria that covered stretches of sand in Scotland's Torridon region. It is thought that the ripples in certain rocks actually represent billion-year-old biosignatures left behind by the first organisms to inhabit the land.

What were the **first land animals** and why did **they move** onto the **dry land**?

The first larger land animals to wander onto land were probably arthropods, such as scorpions and spiders. Many of these creatures have been found in Silurian period rock layers, usually in association with fossils of the oldest-known vascular land plants.

No one truly knows why the first animals moved from the oceans to dry land, but there are plenty of theories. One is that animals wanted to expand their territory, similar to the way many modern animals behave. Another possibility is that as more animals evolved, there would have been a higher demand for a better food source. By adapting to land life—and the "new" food sources on land—these organisms would have a better chance of survival.

When did the **first primitive dinosaurs** appear?

The first primitive dinosaurs appeared about 230 million years ago. They were much smaller, and less fierce, than the *Tyrannosaurus rex* we often think of when someone mentions the word "dinosaur."

How long did it take for **dinosaurs** to evolve from the **first land animals**?

The first larger land animals that would eventually lead to the appearance of dinosaurs evolved around 440 million years ago. Dinosaurs then evolved around 250 million years ago. Thus, it took about 190 million years for dinosaurs to appear after the first land animals. Remember, these numbers are based on the currently known fossil record, and could change if new fossils are found.

GEOLOGIC TIME

What is **geologic time**?

Geologic time is the immense span of time that has elapsed since Earth first formed—almost 4.5 billion years ago—to recent times.

What is the **geologic time scale**?

The geologic time scale is a way of putting Earth's vast history into an orderly fashion, giving a better perspective of events. At the turn of the nineteenth century, William Smith (1769–1839), an English canal engineer, observed that certain types of rocks, along with certain groups of fossils, always occurred in a predictable order in relation to each other. In 1815, he published a map of England and Wales geology, establishing a practical system of stratigraphy, or the study of geologic history layer-by-layer. Simply put, Smith proposed that the lowest rocks in a cliff or quarry are the oldest, while the highest are the youngest.

By observing fossils and rock type in the various layers, it was possible to correlate the rocks at one location with those at other locations. Smith's work, combined with the first discoveries of dinosaur fossils in the early 1800s, led to a framework that scientists still use today to divide Earth's long history into the geologic time scale, with its various, arbitrary divisions of time including eras, periods, and epochs. Established between 1820 and 1870, the time divisions are a relative means of dating; that is, rocks and fossils are dated relative to each other as to which are older and younger. It was not until radiometric dating was invented in the 1920s that absolute dates were applied to rocks and fossils—and to the geologic time scale.

What are the **divisions** of the **geologic time scale**?

The geologic time scale divisions have changed significantly over time, mainly because of new fossil discoveries and better radiometric dating techniques—and it will no doubt continue to change. The following table is a general listing of the geologic time table based on current interpretations of rocks and fossils.

The Geologic Time Scale (in millions of years ago)

Eon	Era	Period	Sub-Period	Epoch
Precambrian 4,500–543	Hadean 4,500–3,800			
	Archaean 3,800–2,500			
	Proterozoic 2,500–543	Paleoproterozoic 2,500–1,600		
		Mesoproterozoic 1,600–900		
		Neoproterozoic 900–543		
Phanerozoic 543 to present	Paleozoic 543–248	Cambrian 543–490		
		Ordovician 490–443		
		Silurian 443–417		
		Devonian 417–354		
		Carboniferous 354–290	Mississippian 354–323	
			Pennsylvanian 323–290	
		Permian 290–248		
	Mesozoic 248–65	Triassic 248–206		
		Jurassic 206–144		

Eon	Era	Period	Sub-Period	Epoch
		Cretaceous 144–65		
	Cenozoic 65 to present	Tertiary 65–1.8	Paleogene 65–23.8	Paleocene 65–54.8
				Eocene 54.8–33.7
				Oligocene 33.7–23.8
			Neogene 23.8–1.8	Miocene 23.8–5.3
				Pliocene 5.3–1.8
		Quaternary 1.8 to present		Pleistocene 1.8 to 10,000 years ago
				Holocene 10,000 years ago to present

How are the **divisions** on the **geologic time scale named**?

Most of the major divisions on the geologic time scale are based on Latin names, or areas in which the rocks were first found. For example, the Carboniferous period gets its name from the Latin words for "carbon-bearing," in reference to the coal-rich rocks found in England; the Jurassic period is named after the Jura Mountains along the border of France and Switzerland. The names of the stages or ages most often depend on city and regions where the rocks were found; this is why division names frequently vary on geologic time scale charts from different countries.

What are the **major time units** used in the **geologic time scale**?

There are five major time units on the geologic time scale. The units are—in order of descending size—eons, eras, periods, epochs, and stages (although some list this division as ages and subages). The eon represents the longest geologic unit on the scale; an era is a division of time smaller than the eon, and is normally subdivided into two or more periods. An epoch is a subdivision of a period; a stage is a subdivision of an epoch.

What do the **divisions** on the geologic time scale **represent**?

The geologic time scale is not an arbitrary listing of Earth's natural history, nor are the divisions merely fanciful. Each boundary between divisions represents a change or an event that delineates it from the other divisions. In most cases, a boundary is drawn to represent a time when a major catastrophe or evolutionary change in animals or plants (including the evolution of specific species) occurred.

Natural erosion clearly reveals the layers of Earth's crust, such as seen here in Badlands National Park in South Dakota. Observing these layers is like taking a trip back in time, with each lower level representing a different time period in the planet's history (iStock).

What is **relative time** in relationship to geologic time?

Relative time is a way to establish the relative age of rocks and fossils. It is based on the location of a rock layer in comparison to the location of other rock layers; that is, it is only relative, not absolute, time. In many cases, rock layers are laid down in order, the older layers being below the younger layers. For example, a fossil found in a higher rock layer is usually younger than a fossil found in a rock layer below it. During the nineteenth century, scientists used this method to date rock layers relative to each other and to establish and construct the first geologic time scale.

What is **absolute time** in relationship to geologic time?

Absolute geologic time is the (approximate) true age of the rock; that is, the absolute time that the rock layer formed. Typically, radiometric techniques, which measure the amount of radioactive decay in rocks, are used to determine absolute time.

When were **radiometric dating techniques** discovered?

The basic principles and techniques of radiometric dating were not discovered until the turn of the twentieth century. In 1896, French physicist Antoine-Henri Becquerel (1852–1908) accidentally discovered radioactivity when a photographic plate left next to some uranium-containing mineral salts blackened, proving that uranium gave off its own energy. In 1902, British physicist Lord Ernest Rutherford

Why do some dates differ on the various geologic time scale charts?

Determining the true age divisions of the past 4.6 billion years for the geologic time scale is not a perfect science. (Determining the date of a rock layer is not as precise as knowing your own age.) In addition, there is often disagreement as to the extent of certain time periods, since rocks and fossils found on different continents vary. Even radiometric dating does not reveal the true age of a rock or mineral because there is always a certain amount of estimation involved.

(1871–1937) collaborated with British chemist Frederic Soddy (1877–1966) to discover that the atoms of radioactive elements are unstable, giving off particles and decaying to more stable forms. These findings led United States chemist Bertram Borden Boltwood (1870–1927) to argue that, by knowing the decay rate of uranium and thorium into lead, the dating of rock would be possible. In 1905, Boltwood and John William Strutt dated various rocks, obtaining ages of 400 to 2,000 million years for various rock samples and proving such dating could be done.

Who first developed an **absolute geologic time scale** using **radiometric dating**?

In 1911, British geologist Arthur Holmes (1890–1965) began to formulate a geologic time scale based on absolute time, using the uranium-lead dating method to determine the age of rocks. In 1913, he published *The Age of Earth,* in which he outlined how radioactive decay methods, in conjunction with geological data, could be used to construct an absolute geologic time scale. In 1927, Holmes estimated that the age of Earth's crust, based on his radiometric techniques, is approximately 3.6 billion years old.

TIME PERIODS

What is the **Pre-Cambrian era**?

The Pre-Cambrian era represents the time of Earth's beginning to just before the big explosion of life in the oceans—from about 4.54 billion to about 543 million years ago. During this time, Earth was cooling, developing its oceans, and building the continental crust; in addition, scientists believe that life began during the early part of the Pre-Cambrian. The following lists one interpretation of three Pre-Cambrian divisions, the approximate dates, and major evolutionary events during these times:

Hadean—4.5 to 3.8 billion years ago, the time when Earth was forming in the early solar system.

Archaean—3.8 billion years to 2.5 billion years ago, the oldest bacteria evolved.

Proterozoic Era—2.5 billion to 543 million years ago, a time in which multi-celled eukaryotic (a cell with a definitive nucleus) evolved—in other words, animals.

Why do scientists believe that several **ice ages** occurred during the **late Pre-Cambrian era**?

Chemical and isotopic analysis of rocks found in Africa show that Earth may have gone through at least four ice ages between 750 and 570 million years ago. These were very deep ice ages, essentially turning Earth into a "snowball planet." From the evidence to date, some scientists think the oceans were covered with ice almost 300 feet (91 meters) deep, and the land was completely dry and barren of life.

Some scientists believe the Pre-Cambrian ice ages may have been caused by Earth's tilt toward the Sun. The planet may have been tilted at a much larger angle—upwards of 55 degrees—than today's angle of 23.5 degrees. This large degree of tilt meant that the polar areas received most of the Sun's warmth, keeping them ice-free. But the areas around the equator would have been colder, allowing glaciers to form. If this was true, the buildup and melting of the glaciers around the equator during the Pre-Cambrian era may have created enough force to move the planet's axis to its modern position. Some scientists have equated this process to repeatedly pushing on a swing at just the right moment in its movement, adding energy to make it go higher. The influence of the alternating advance and retreat of the glaciers could have caused the axis to straighten to its present angle.

Some scientists believe the "heroes" that thawed the snowball planet and paved the way for an explosion of life were none other than volcanoes. As these surface blisters erupted toward the end of the Pre-Cambrian era, they sent massive amounts of carbon dioxide into the atmosphere, an increase of approximately 350 times its present concentration. This increase trapped re-radiating solar energy, warming the planet as it created a super greenhouse effect. The temperatures rose enough to melt the ice-covered oceans and end the ice age.

Why was the **"Cambrian Explosion,"** also called the **"evolutionary big bang"**?

Just after the end of the Pre-Cambrian era, about 543 million years ago (during the Cambrian period), a great burst of evolutionary activity began in the world's oceans. Based on the fossil record of the Cambrian period, scientists estimate that the number of orders of animals doubled roughly every 12 million years. At this time, too, most of the modern phyla of animals began to appear in the fossil record.

For some reason, new animals appeared at breakneck speed, geologically speaking, filling the oceans with life. No one really knows why the animals started to

Trilobites were one of the most successful creatures to ever inhabit our planet, with some 15,000 species that survived from the Early Cambrian through the Permian periods (iStock).

appear, and scientists have suggested theories ranging from a change in climate to an overall natural threshold reached. For example, some scientists believe temperature or oxygen levels reached a point that allowed the proliferation of organisms.

By looking at the genes preserved in and common to modern animals, researchers are trying to determine a possible cause. One study found that an ancient common ancestor—a worm-like animal from which most of the world's animals were derived—had special genetic machinery that was so successful that it survives to this day. These genes, used to grow appendages (arms, legs, claws, fins, and antennas), were operational at least 600 million years ago. With appendages, animals swam faster, grabbed tighter, and fought with greater efficiency, and, thus, they could eventually dominate the globe.

What were **trilobites** and when did they exist?

Trilobites were some of the most successful creatures to ever live on Earth—and finding one of their fossils is one of the joys of modern fossil collecting. These hard-shelled, segmented animals lived for hundreds of millions of years in the oceans; they are called "three-lobed" because of their head, thorax, and tail. They were one of the first arthropods (animals with jointed legs), comprised over 15,000 different species, and ranged in size from about a half inch (3 millimeters) to 2.3 feet (70 centimeters) long.

Trilobites first appeared in the Early Cambrian (during what is called the "Cambrian Explosion"); they increased in diversity in the Devonian period; and their numbers decreased in the Silurian (probably due to the appearance of sharks and other predators). They became extinct after the Permian mass extinction. Although they did not exist during the time of the dinosaurs, scientists respect this species for its tenacity and abundance.

What **theories** attempt to explain why the **Cambrian Explosion** occurred?

In the past, paleontologists have offered many theories as to why the great blossoming of life, known as the Cambrian Explosion, occurred. Some point to the mass extinction of Ediacara organisms (the earliest known complex multi-cellular organisms); another theory states that the development of eyes in animals changed the predator-prey dynamics; and yet another theory suggests that an increase in size of many organisms accelerated diversity.

There is also one study that places the blame, or credit, for this evolutionary explosion on our planet itself—and although not everyone agrees, it is an interesting theory. More than 500 million years ago, shifting masses in the interior mantle of Earth essentially unbalanced the planet, or "tipped" it, causing the entire surface to reorient itself in an effort to become balanced again. In a process called "true polar wander," the ancestral North America moved from near the South Pole up to the region of the equator; the large continent of Gondwanaland (made up of South America, Antarctica, Australia, India, and Africa) traveled all the way across the Southern Hemisphere. This movement all happened at more than twice the rate of continental drift found in the normal process of plate tectonics today.

Evidence for this theory comes from Earth itself. During the formation of rocks, the minerals inside naturally align themselves with the existing magnetic field of the planet. By studying the orientation of grains in the minerals, scientists can determine the position of ancient continents relative to the magnetic north pole, which almost always lies close to Earth's axis of rotation. When the positions of the continents were plotted using this data, scientists found that there was a major movement of the continents within a relatively short period of time around the Cambrian period. The data showed that ancestral North America moved to the equator between 540 and 515 million years ago, while Gondwanaland shifted between 535 and 500 million years ago.

What are the **Paleozoic, Mesozoic,** and **Cenozoic** eras?

The divisions between the eras on the geologic time scale represent major changes on Earth. The division between the Pre-Cambrian and Paleozoic, about 543 million years ago, represents an increase of life on Earth. The division between the Paleozoic and Mesozoic represents a major decrease in plant and animal species (called an extinction) about 250 million years ago. It is also called the "Permian Extinction," or the "Great Dying," in which up to 90 percent of all species died out. The division between the Mesozoic and Cenozoic, about 65 million years ago, also represents a major extinction of plant and animal species, including the dinosaurs.

This extinction was not as extensive as the Great Dying: only about 50 percent of all species died out at this time.

What are the **divisions** of the **Mesozoic era**?

The Mesozoic era, often referred to as the "age of the reptiles" or the "age of the dinosaurs" (even though dinosaurs did not evolve until well into the Mesozoic), lasted from approximately 250 to 65 million years ago. It is divided into three periods: the Triassic, Jurassic, and Cretaceous.

What are the **more recent time divisions** on the **geologic time scale**?

The Cenozoic era is divided into the Tertiary and Quaternary (or Anthropogene) periods. The Quaternary is further divided into the Pleistocene epoch, a period of advances and retreats of huge ice sheets; and the Holocene epoch, or recent times, which began about 10,000 years ago.

FIRST FOSSILS

What is a **fossil**?

The remains of plants and animals that have been preserved in the earth, close to their original shape, are called fossils. This word comes from the Latin *fossilis,* meaning "something dug up." The different types of fossils depend on the remains and conditions present at the time the organism died. Fossils may be formed from the hard parts of an organism, such as teeth, shells, bones, or wood; they may also be unchanged from their original features, the entire organism having been replaced by minerals such as calcite or pyrite. Animals and plants have also been preserved in other materials besides stone, including ice, tar, peat, and the resin of ancient trees.

Fossils of single-celled organisms have been recovered from rocks as old as 3.8 billion years. Animal fossils first appeared in rocks dating back over one billion years ago. The occurrence of fossils in unusual places, such as dinosaur fossils in Antarctica and fish fossils on the Siberian steppes, is due to the shifting of the continental plates that make up Earth's crust, and environmental changes over time, such as an ice age. The best explanation of dinosaurs in Antarctica is not that they evolved there, but that Antarctica was once part of a much larger landmass with which it shared many life forms.

How does a **fossil form**?

There are a number of ways a fossil forms, depending on the type of remains and the environment present. In general, the process for most fossils is much the same: the hard parts of animals, such as bones, teeth, and shells, as well as the seeds or woody parts of plants, are covered by sediment, such as sand or mud. Over millions of years, more and more layers of sediment accumulate, burying these remains deep

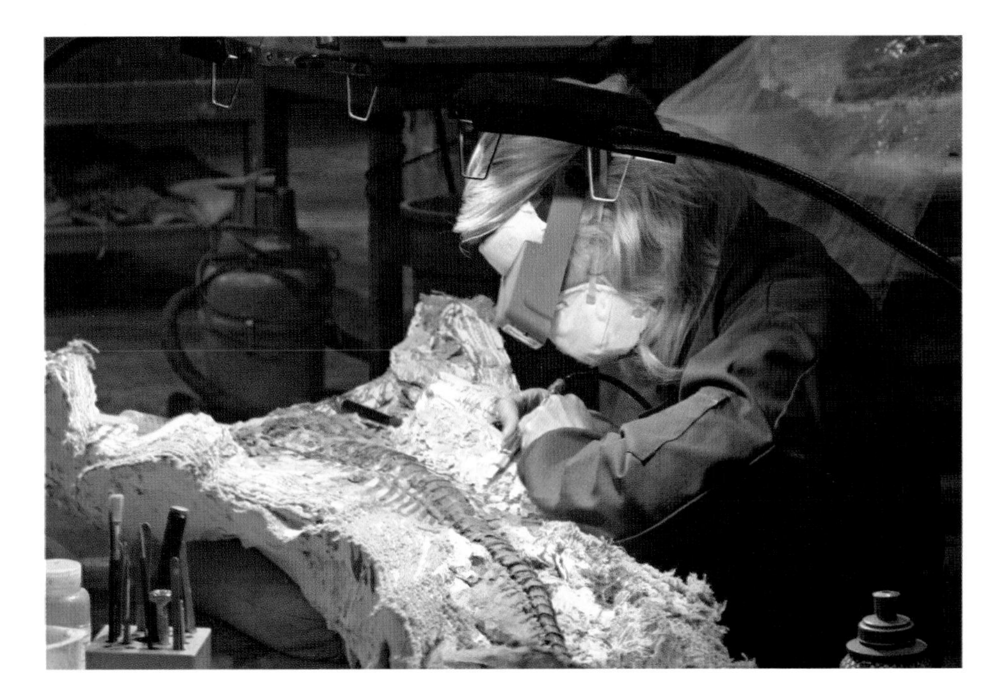

Uncovering a fossil from the surrounding rock is meticulous work that can take hundreds of hours because paleontologists do not wish to damage an artifact that took millions of years to form (iStock).

within the earth. The sediment eventually turns to stone, and often the remains are chemically altered by mineralization, becoming a form of stone themselves (these are the type of fossils often viewed as the recreated dinosaur skeletons seen in many museums). The same process also produces petrified wood, coprolites (petrified excrement), molds, casts, imprints, and trace fossils.

Most fossils are found in sedimentary rocks—those rocks produced by the accumulation of sediment such as sand or mud. Wind and other weathering conditions wash away sediment on land, depositing it in bodies of water. For this reason, fossils of sea creatures are more common than those of land creatures. Land animals and plants that have been preserved are found mostly in sediment of calm lakes, rivers, and estuaries.

A fossil may also consist of unaltered original material. Bones and teeth are commonly preserved in this way. However, far more often the pores of bone and teeth are filled with minerals in a process called permineralization (what many have called petrifying). Circulating ground water carries silica or calcium carbonate (and sometimes other minerals, such as pyrite) that fill the pores. What remains is, in essence, a duplicate of the original bone or other organic material.

How **likely** is it that an **organism becomes a fossil**?

Not all organisms survive to become fossils, and the chance of a living organism becoming a fossil is generally very low. Many organisms completely decay away or are chewed apart by other animals. Because of this, some scientists estimate that although billions of flora and fauna have lived on Earth, very few survived into fossil form. The fossils we do find represent only a fraction of the animals and plants that ever lived.

An organism has the best chance to become a fossil if it is quickly covered by moist sediment after death, protecting the decaying organisms from predators, scavengers, and bacteria. The soft parts of the organisms (such as skin, membranes, tissues, and organs) quickly decay, leaving behind bones and teeth. The majority of found fossils date back no farther than almost 500 million years ago, when organisms first began to develop skeletons and other hard parts.

The following are the steps to fossilization, using a dinosaur as an example. This outline shows how difficult it is for a dinosaur to become a fossil:

Scavenging and decay—When a dinosaur died, it did not take long for scavengers to remove the soft flesh parts of its body. Those parts that were not eaten decayed at a fast or slow rate, depending on the prevailing climate. In any case, within a short amount of time only a skeleton would remain. But even the remaining hard body parts were not impervious to change. They were often weathered by the action of wind, water, sunlight, and chemicals in the surroundings, rounding the bones or reducing them to small pieces.

Location—If the dinosaur's skeleton was in an area in which rapid burial did not take place, then the chances of fossilization were slim. The bones would break and scatter, often moved by the action of changing river courses or flash floods. But occasionally, this transport increased the chance of fossilization, moving the bones to a better area for preservation, such as a sandbank in a river.

Burial—The most crucial step in becoming a fossil is burial. The sooner the burial of the dinosaur bones, the better the chance of a good fossil being created. If the bones were covered by mud or sand, whether before or after transport, then the amount of further damage would have lessened; in addition, the exposure to oxygen was less, thus reducing additional decay of the dinosaur bones. Some damage might still have occurred, however, primarily from the pressure created by the increasing amount of sediment on top of the bones, or even from acidic chemicals that dissolved into the sediment.

Fossilization—The fourth step is the actual process of fossilization itself. Here, the sediments surrounding the fossil slowly turn to stone by the action of pressure of the overlying sediment layers and loss of water. Eventually, the grains become cemented together into the hard structure we call rock. The dinosaur bones fossilized, as the spaces in the bone structures fill with minerals, such as calcite (calcium carbonate), or other iron-containing minerals; or the actual mineral component of the bone itself, apatite (calcium phosphate), may have recrystallized.

Exposure—Lastly, deeply buried dinosaur bones must be exposed on the surface where they can be discovered. This involves the uplift of the bone-containing sedimentary rock to the surface, where erosion by wind and water expose the fos-

Why are there gaps in the fossil records?

Gaps in the fossil records—eras or evolutionary stages that are "missing" from the known collection of fossils—are most often the result of erosion. This geologic process erodes away layers of rock and embedded fossils, usually by the action of wind, water, and ice. Gaps in fossil records can also be caused by mountain uplift, which destroys fossils, and volcanic activity, which can bury fossil evidence with hot magma rock that physically changes the rock, and thus fossils.

silized skeleton. If the bones are not found in time, the action of the wind and water can destroy the precious record of the ancient species.

How do scientists determine the **age** of **fossils**?

A number of methods are used today to date fossils. Most of the methods are indirect—meaning that the age of the soil or rock in which the fossils are found are dated, not the fossils themselves. The most common way to ascertain the age of a fossil is by determining where it is found in rock layers. In many cases, the age of the rock can be determined by other fossils within that rock. If this is not possible, certain analytical techniques are often used to determine the date of the rock layer.

One of the basic ways to determine the age of rock is through the use of radioactivity. For example, radioactivity within Earth continuously bombards the atoms in minerals, exciting electrons that become trapped in the crystals' structures. Using this knowledge, scientists use certain radiometric techniques to determine the age of the minerals, including electron spin resonance and thermoluminescence. By determining the number of excited electrons present in the minerals—and comparing it to known data that represents the actual rate of increase of similar excited electrons—the time it took for the amount of excited electrons to accumulate can be calculated. In turn, this data can be used to determine the age of the rock and the fossils within the rock.

There are other methods for determining fossil age. For example, uranium-series dating measures the amount of thorium-230 present in limestone deposits. Limestone deposits form with uranium present and almost no thorium. Because scientists know the decay rate of uranium into thorium-230, the age of the limestone rocks, and the fossils found in them, can be calculated from the amount of thorium-230 found within a particular limestone rock.

What are **molds** and **casts**?

Molds and casts are types of fossils. After burial, a plant or animal often decays, leaving only an impression of its hard parts (and less often, soft parts) as a hollow mold

Fossils are not always bones. These fish fossils are not actually bones, but rather imprints the fish made in the soil. (iStock).

in the rock. If the mold is filled with sediment, it can often harden, forming a corresponding cast.

What are **trace fossils**?

Not all fossils are hardened bones and teeth, or molds and casts. There are also fossils that are merely evidence that creatures once crawled, walked, hopped, burrowed, or ran across the land. Trace fossils are just that: the traces of a creature left behind, usually in soft sediment like sand or mud. For example, small animals bored branching tunnels in the mud of a lake bed in search of food; and dinosaurs hunted for meals along a river bank, leaving their footprints in the soft sand. Similar to the fossil formation of hard parts, the footprints and tunnels were filled in by sediment, then buried by layers of more sediment over millions of years, eventually solidifying. Today we see the results of this long-ago activity as trace fossils. Many originators of trace fossils are unidentifiable—in other words, there are no hard fossils of the creatures left in the area, just their tracks. Some of the most famous trace fossils are those of dinosaurs tracks (for example, in Culpepper, Virginia, and near Golden, Colorado), and human-like footprints (for example, in east Africa), which were all found in hardened sediment.

There is a difference between tracks and trails, too: tracks are generally the traces of distinct footprints, whereas trails may have been produced by an animal dragging its feet or some other appendage as it moved. Tracks, therefore, are more distinctive, and different animals can be distinguished by their own particular footprints. Trails can seldom be associated with a particular animal.

What can some **trace fossils** of dinosaur tracks **tell us**?

Numerous fossilized dinosaur footprints, called a trackway, indicate much about dinosaurs' speeds. One such trackway is located north of Flagstaff, Arizona, on the

Navajo Reservation. This site was first discovered by Barnum Brown of the American Museum of Natural History in the 1930s, but had been lost until recently. It includes an example of a running dinosaur that left tracks with an 8-foot (2.4-meter) space between the right and left prints; from these prints, scientists calculated that the dinosaur ran at speeds of 14.5 miles (23.3 kilometers) per hour—one of the faster dinosaurs known. The record for fastest dinosaur, however, is presently held by a Jurassic carnivore that left a 16-foot (5-meter) gap between the right and left tracks in a Glen Rose, Texas, trackway. The calculated speed of this dinosaur was about 26.5 miles (42.8 kilometers) per hour, much faster than the speediest human.

DINOSAUR FOSSILS

How do paleontologists **identify species of dinosaurs** from other fossils?

One of the best ways to identify dinosaur fossil bones is by size, as many of the bones are huge. For example, the upper leg bone, or femur, of an adult *Apatosaurus* often measures over 6 feet (1.8 meters) long.

But size is not everything, as many dinosaurs were the same size as a chicken or cat. The way scientists detect the differences between dinosaurs and other animal species is by the construction and orientation of their bones, including heads, tails, and hipbones. In addition, dinosaur fossils are often found in association with other dinosaurs at a site. Many times these fossils represent dinosaurs—from meat-eaters to plant-eaters—that gathered together along the shore of a lake or ocean. The dinosaurs were all searching for food along the banks of the water, a place that would attract many animals and plants.

Of course, not all dinosaurs are found in the conventional way. In 1998, an amateur fossil collector saw the movie *Jurassic Park,* and recognized that a fossil he had (which he thought was a bird) was actually a dinosaur. The specimen was found in Italy and measures only 9.5 inches (24 centimeters) long; scientists now know it was a young dinosaur (a theropod named *Scipionyx samniticus*) that had just hatched before it died. In this case, the dinosaur remains probably washed into oxygen-starved waters, where it was quickly buried. This may be one of the most important dinosaur fossils ever found, as many of its soft body parts were preserved, including the intestines, muscle fibers, and what appears to be the liver.

When was the first **dinosaur bone** collected and described?

The fossilized bones of dinosaurs have probably been found throughout human history, but for much of this time people did not realize what they were. Therefore, no records or descriptions were kept until fairly recently. References to fossilized sharks' teeth and shells are recorded from the European medieval period, but because they believed that no animal or plant made by God could become extinct, they explained them in other ways. For example, many of the fossils were interpreted as the remains of modern species as opposed to ancient, extinct species; others were thought of as merely pebbles that resembled the remains of animal and plant species.

The first recorded description of a dinosaur bone was made in 1676 by Robert Plot (1640–1696), a professor of chemistry at the University of Oxford, England, in his book *The Natural History of Oxfordshire*. Although he correctly determined that it was a broken piece of a giant bone, Plot did not know the bone came from a dinosaur. Instead, he felt it belonged to a giant man or woman, citing mythical, historical, and biblical sources. In 1763, the same bone fragment was named *Scrotum humanum,* by R. Brookes, to describe its appearance, but the name never gained wide, or serious, acceptance. Based on Plot's illustration, modern scientists believe the bone fragment is actually the lower end of a thigh bone from a *Megalosaurus,* a meat-eating dinosaur from the Middle Jurassic period that roamed the area now known as Oxfordshire.

In 1787, American physician Caspar Wistar (1761–1818) and Timothy Matlack (better known as a statesman and patriot during the American Revolution; 1730–1829) discovered a large fossil bone in the state of New Jersey. Although they reported their finding, it was ignored and unverified; it may have been the first dinosaur bone ever collected in North America.

What are some of the **oldest dinosaur fossils** found to date?

There are several dinosaur fossils claimed to be the oldest ever found, a claim many scientists have made in recent years. The oldest dinosaur bones and skulls found to date may be of a small genus of meat-eating animal measuring about 6.5 feet (2 meters) in length, but some scientists believe it was another type of reptile called a thecodont. The fossils were discovered in Brazil and are thought to be 235 to 240 million years old.

Previously, an *Eoraptor,* a 228 million-year-old meat-eating dinosaur, was found in Argentina. More recently, paleontologists also discovered the fossilized bones of three prosauropods, plant-eating dinosaurs that lived approximately 220 million years ago in Santa Maria, an area of southern Brazil. And finally, another claim to the oldest-known "dinosaurs" (or precursors) were prosauropods from Madagascar thought to be 230 million years old. The *Eoraptor* was dated using radio-isotope analysis, whereas index fossils—or the use of surrounding fossils to date a "dinosaur"—were used to date the fossil in Madagascar.

EVOLUTION OF THE DINOSAURS

What is a **dinosaur**?

Dinosaur is a term used to describe certain types of animals that lived during the Mesozoic era in geologic history. It is difficult to generalize about dinosaurs, but two things are definitely agreed upon. They were, in general, the largest creatures to ever walk on Earth, even though there were many smaller species of dinosaurs. In addition, these animals were some of the most successful organisms that ever lived, existing as species for at least 160 million years.

What does the term **dinosaur mean**?

Dinosaur comes from the term *dinosauria,* which is a combination of the Greek words *deinos* and *sauros*. It means "terrible reptiles" or "terrible lizards." The term was invented by the well-known British anatomist Sir Richard Owen (1804–1892). He coined the term in 1842 to describe the 175-million-year-old fossil remains of two groups of giant reptiles that corresponded to no known living creatures. In 1854, Owen prepared one of the first dinosaur exhibits for display at the Crystal Palace in London, England.

DINOSAUR ANCESTORS

How did **life evolve after** the early **one-celled organisms**?

Over hundreds of millions of years after the evolution of single-celled organisms, the oceans abounded with a huge variety of life. The first soft-bodied animals, such as worms and jellyfish, evolved toward the close of the Pre-Cambrian (also seen as Precambrian) era roughly 600 million years ago; the first animals with hard parts, such as shelled mollusks, evolved during the first period after the Pre-Cambrian known as the Cambrian period of the Paleozoic era.

What are **vertebrates**?

The first vertebrates, or animals with backbones, evolved during the late Cambrian to early Ordovician periods as jawless freshwater fish that looked much like today's hagfish and lampreys. By the Devonian period (the "age of fishes"), jawed and armored fishes dominated the oceans. Around 380 million years ago, a line of fish with bony skeletons developed air-breathing lungs and "limbs" strong enough to support them. These were the precursors to the amphibians, creatures that made their first move toward land probably in response to the spread of plants to land around the early Silurian period.

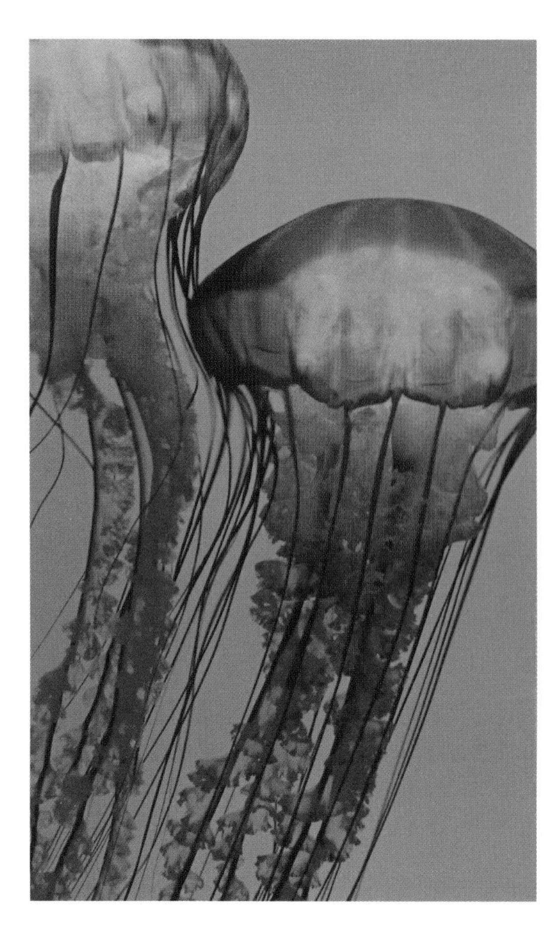

Still around today, jellyfish were some of Earth's first soft-bodied life forms (iStock).

What were the **early amphibians** and when did they live?

Amphibians were the first air-breathing land vertebrates, evolving from lobe-finned fish and primitive tetrapods. Tetrapods were animals with fish-like heads and tails, and with limbs that were little more than jointed, lobed fins. They evolved around 340 million years ago during the late Devonian period. So far, the oldest of these pre-amphibian fossils have been dated at approximately 360 million years old. These animals could do something that no fish could do: breathe air. The changeover from gills to lungs came during the early larval stage of the amphibian. These early amphibians were direct descendants from the early fish, and represent an important transitional stage from water-dwelling to land-dwelling animals. They were also the first vertebrates to eventually evolve true legs, tongues, ears, and voice boxes.

The word "amphibian" comes from the Greek *amphi,* meaning "both," and *bios,* meaning "life" (sometimes translated as "living a double life"). The name signifies that these animals could live both in and out of the water.

The Carboniferous period of the Paleozoic era, from approximately 360 to 280 million years ago, brought about a proliferation of amphibians, as did the Permian, from about 280 to 248 million years ago. Much of the world's climate during both time periods was warm and humid, with many swamps, marshes, and lakes dominating the landscape—a perfect environment for the water needs of the amphibians. In some texts, the Carboniferous or the Permian periods are called the "age of amphibians" (although many scientists agree that reptiles began to take over the amphibians' domain during the Permian).

What were some of the **earliest species** that **led to amphibians**?

Several fossils of pre-amphibian species have been uncovered. For example, one of the earliest known steps to amphibians is the *Acanthostega,* which evolved about 360 million years ago during late Devonian. It was one of the first to have feet; it also had toes (eight per limb), no fin rays, a large, load-bearing pelvis, and may have retained its gills into adulthood. Another early amphibian fossil is an intermediate between fish and amphibian: the *Ichthyostega,* with the best fossils having been uncovered in Greenland. This early amphibian lived in the swamps of the Late Devonian period, enjoying mild, warm climates. By this time, too, insects had evolved on land, providing food for the slow-moving amphibian. The *Ichthyostega* was a three-foot- (one-meter-) long animal with four limbs and a fin on its tail—a combination of amphibian and fish features that allowed it to swim and to climb on land.

What were some of the **problems amphibians faced** in moving from **water to land**?

The early amphibians' main problem was support. In the water, a body is virtually "weightless" because it is supported by the buoyancy of water. But on land, an amphibian's body had to be held up from the ground, and the internal organs needed to be protected from being crushed by gravity; thus, a strong ribcage was essential. The backbone, ligaments, and muscles also had to be strong, supporting not only the weight of the body between the front and hind legs, but also the head. The limbs and limb muscles had to change design to allow walking. Hind limbs became attached to a supportive pelvis, and the skeleton as a whole was made stronger.

Another problem was adapting to breathing on land. Early amphibians had to modify their respiratory system (changing from gills to lungs), as lungs took over more and more of the breathing. The reproductive system, water balance, and senses also had to adapt to the new life in and out of the water. For example, the first amphibians probably spent much of their time in the water, giving birth to totally aquatic young (tadpoles) that would eventually be able to live both in and out of water. Amphibians' water dependency adapted to allow them to live out of water as long as they at least stayed damp. Their senses also had to adapt—their sight, smell, and hearing taking on more important roles. For instance, amphibian eardrums developed to enable the semi-land dwelling animals to hear sounds in the air. Their eyes had to modify in order to see in air instead of water; protective eyelids developed; and tear ducts evolved, allowing their eyes to be continually moistened with tears.

Just like this paddle tail newt, ancient amphibians survived on both land and sea; they were the first animals to survive for extended periods outside the water (iStock).

Even after all these changes, amphibians still were tied to ponds, lakes, or the edges of the oceans, especially since the eggs still had to be laid and hatched in water. Evolution did not change the amphibians too much—modern amphibians are still tied to water.

When did **reptiles** first **evolve** from amphibians?

It is thought that during the Carboniferous period, a group of amphibians gave rise to the reptiles. The first reptiles were small, lizard-sized animals, but they had many differences from their amphibian ancestors, including waterproof skin and thick-shelled eggs. This made it unnecessary for the reptiles to stay near water, to keep moist, or to lay their eggs in water. In fact, the evolution of the reptiles, geologically speaking, occurred very rapidly. Within about 40 million years, the reptiles had produced thousands of different species.

What was one of the **most important changes** that enabled **reptiles** to become true **land-dwelling** animals?

The development of the amniote egg freed the reptiles completely from life in the water by allowing them to fully reproduce on land. Unlike the young of amphibians, who had to go through a larval stage in the water before metamorphosing into an adult, the amniote egg acted as a sort of "private pond" for the young reptiles.

The egg itself had a hard shell, which contained numerous small pores. These pores allowed air to enter, but the shell prevented the inside from drying up as long as the surroundings were humid. The eggs were fertilized inside the mother's body before being laid. There were three very thin bags inside the shell itself, each of which had a specific function. The first bag held the developing young and a liquid (which took the place of the pond or stream); this area was called the amnion, from which the egg gets its name. The next bag contained the yolk, the source of food for the developing embryo. The third bag was in contact with the air diffusing in through the shell. Thus, the young reptiles had food, air, protection from predators, and an aquatic environment in which to grow. The young would eventually hatch into a miniature version of its parents and was able to fend for itself. Because of the egg, the reptiles no longer had to have a source of water to reproduce and could spread out, populating the land as well as hunting for prey well away from water.

What were some of the **earliest known reptiles**?

Two of the earliest known reptiles, the *Hylonomus* and *Paleothyris,* both descended from amphibians during the Middle Carboniferous period of the Paleozoic era.

What amphibians are living today?

Names of modern amphibians are familiar to us: frogs, toads, salamanders, newts. They represent the descendants of groups that did not become extinct at the end of the Mesozoic era (when dinosaurs died out). Of the modern amphibians, the newts and salamanders are probably the most similar to the early amphibians, although the modern-day versions are much smaller.

The vertebrate class Amphibia today includes about 3,500 species in three orders: frogs and toads (order Anura), salamanders and newts (order Caudata), and caecilians (order Gymnophiona). There is, however, a much larger number of extinct species of amphibians: this ancient group of animals were the first vertebrates to begin exploiting terrestrial environments, where they became prey for other species.

The best evidence of the change from amphibian to reptile was the early reptiles' high skulls—evidence of additional jaw muscles—and thicker egg shells. The *Hylonomus* still claims the prize (so far) as the oldest-known reptile and lived about 315 million years ago. The *Paleothyris* evolved about 300 million years ago. The fossils of both these reptiles were found near Nova Scotia, Canada, in ancient tree stumps. Apparently, the animals fell into tree stumps in pursuit of insects or worms. There they were trapped and eventually died.

Why did the **reptiles dominate** during the **Mesozoic** but not the amphibians?

Besides the ability to not depend on water as much as amphibians, there are probably two main reasons why reptiles became dominant in the Mesozoic. First, reptiles developed adaptations in their skeletal structure, allowing them to move much quicker than amphibians. Second, during the Permian period the climate became hotter and drier, and many water sources disappeared. The reptiles' new adaptations—from the development of scales to retain water to eggs that could survive without staying in water—allowed them to thrive where amphibians could not.

Did some **reptiles return** to the **oceans**?

Yes, as the reptiles spread out over the land, some of them returned to the water. Over a period of time, they evolved and adapted to the water again. Their legs gradually evolved back into fins and flippers; eyes adapted to seeing underwater; and bodies became streamlined for better speed in the water. In addition, they could no longer lay their eggs on land. Thus, they evolved a way of producing living young within their bodies, a process called ovoviviparous. The Ichthyosaurs, or "fish lizards," were the most fishlike true reptiles.

Though it looks much like modern fish do today, the Ichthyosaur was still a reptile, and one of the first true reptiles to live in the water exclusively (iStock).

How are **reptiles grouped**?

During the 100 million years after the first reptiles appeared, various reptile lines continued to evolve. Today, it is difficult to find agreement about reptile classification. In most cases, they are divided into four living orders (the others have died out over time):

Crocodilia—Crocodiles, alligators, gharials, and caimans, comprising 23 known species.

Squamata—Lizards, snakes, and the worm lizards, or amphisbaenids, which make up about 7,900 species.

Testudines—Turtles and tortoises, which includes about 300 species.

Sphenodontia—The endangered tuatara, which can only be found in New Zealand and consists of two species.

There is also another older method of grouping reptiles: subclasses according to the positioning of the temporal fenestrae, or the openings in the sides of the skull behind the eyes: the anapsids, synapsids, diapsids, and euryapsids. The anapsids had no openings in the skull and eventually evolved into today's turtles and tortoises. The synapsids, or "same hole," had a low skull opening, and were once thought to be the ancestors of modern mammals (and are now not considered to be true reptiles). It was the animals of the diapsid line, or "two skull openings," that eventually gave rise to the dinosaurs. Another debatable line is the euryapsids, characterized by a single opening on the side of the skull, which are now usually included with the diapsids.

DINOSAURS FIRST APPEAR

How did the **reptiles give rise** to the **dinosaurs**?

It was the diapsid group of the reptiles that eventually produced the dinosaurs. The jaw muscles in these reptiles were attached to the two openings on each side of the skull, giving their jaws better leverage and strength. Sometime in the Permian period, the diapsid line branched into two groups: lepidosaurs and archosaurs. The lepidosaur group evolved into today's lizards and snakes; the archosaurs eventually gave rise to the dinosaurs.

What were **early archosaurs** like?

One of the first archosaurs—and probably typical of many archosaurs—was the big-headed *Shansisuchus,* a Middle Triassic period creature that lived in what is now China about 220 million years ago. It was about 7 feet (2.2 meters) long and had long back legs and short front legs.

What were **other archosaurs** like?

Archosaurs varied in many ways. Some, like the 1-foot- (30-centimeter-) long *Gracilisuchus* of the Middle Triassic could run on their hind legs over short distances; others, like the 12-foot- (4-meter-) long *Chasmatosaurus,* were heavy carnivores that walked on all fours. The 2-foot- (0.7-meter-) long *Lagerpeton* had very peculiar hind limbs: its feet had an elongated fourth toe that may have been used for perching.

What are **thecodonts**?

Thecodonts ("socket teeth") were once thought to be a group living from the Late Permian to the end of the Triassic period—a division within the archosaurs that eventually gave rise to the dinosaurs, and perhaps crocodiles, birds, and pterosaurs. It is now considered an obsolete term—and there are definite reasons why.

Just like many fields of science, the study of early dinosaurs constantly changes—not to mention that differences in opinion abound. By the mid-1980s, many scientists proposed that there was *not* a group called the thecodonts that evolved from the archosaurs. In fact, most scientists now believe Thecodontia as a group does not exist. Scientists cite that some of the animals listed in this group are more closely related to the crocodiles, some to dinosaurs, and some to the entire group of archosaurs. Other scientists believe that while thecodonts may not be a true group, the term was a handy way to describe certain creatures with socketed teeth within the archosaurs.

What was the **next stage** in the **evolution** towards true dinosaurs?

As time went on, another phase of dinosaur evolution took place. The animals' skeletal structure changed, especially the hips, which gave many of the dinosaurs the ability to run on two legs. *Euparkeria* was a small lizard-like reptile that lived on land and walked on all fours, but it could run on two legs when in a hurry. Further along in time was *Ornithosuchus,* or "bird crocodile," which was a two-legged predator. Its front limbs were too small to use for walking on all fours, and its thighs were nearly vertical. From this evidence, it appears that *Ornithosuchus* walked only on its hind legs.

How were the dinosaurs unique among the reptiles?

Dinosaurs, unlike other reptiles, had their legs tucked in underneath their bodies. This gave them the ability to run and walk very efficiently, eventually leading to some species becoming totally bipedal (two-footed). They also had a keener sense of smell, sight, and hearing, unlike the amphibians and primitive reptiles.

What are some of the **earliest** known **primitive dinosaurs**?

Two of the earliest known primitive dinosaurs were both fast-running carnivores: the *Eoraptor,* or "dawn hunter," a small, three-foot- (one-meter-) long dinosaur that weighed about 22 pounds (10 kilograms); and the *Herrerasaurus,* measuring from 9 to 18 feet (3 to 6 meters) long. Both lived approximately 230 million years ago in the area known today as Argentina. Still another earlier dinosaur was the *Staurikosaurus,* a Late Triassic (around 225 million years ago) carnivore found in Brazil. It was about 6.5 feet (2 meters) long, with a fully upright gait that allowed for speed. After the evolution of these early specimens, other dinosaurs evolved quickly, becoming more and more diverse, and reachng out into all ecological niches.

What were **early carnivorous** and **herbivorous dinosaurs** like?

The earliest carnivorous dinosaurs, or meat-eaters, came in many different shapes and sizes. It is thought that the 20-foot- (6-meter-) long, Early Jurassic *Dilophosaurus* was a typical carnivore: This dinosaur had strong hind legs, but short, weak forelimbs. It also had thin parallel ridges on its forehead, which could have acted as radiators to control temperature, or as decoration, possibly for territorial or mating displays. The earliest herbivorous dinosaurs, or plant-eaters, also came in many different shapes and sizes. One typical herbivore was the Early Jurassic *Heterodontosaurus,* a small, turkey-sized dinosaur that had sharp incisors, canine-like tusks, and grinding teeth for chewing plants.

How are **dinosaurs classified**?

All animals (and plants) fall into a modern classification—a system first developed by Swedish naturalist Carolus Linnaeus (Carl von Linné, 1707–1778). In the standard Linnean system of animal classification, the hierarchy to dinosaurs is as follows: Animalia (kingdom); Chordata (phylum); Reptilia (class); and Dinosauria (infraclass).

There are even more divisions, and the list seems to get longer and longer every time another dinosaur fossil is found. But in general, they are often divided into two main groups, based historically on their hipbone structure. Those with hips that had the two lower bones pointing in opposite directions, with the pubis bone pointing forward are called saurischian, or lizard-hipped dinosaurs. Those with hips that had the two lower bones lying together behind the back legs, and the pubis bone pointing backward are called ornithischian, or bird-hipped dinosaurs. *Tyrannosaurus rex* is an example of a lizard-hipped dinosaur; while the *Iguanodon* is an example of a bird-hipped dinosaur.

How has the **definition** of **dinosaur changed** over the years?

In 1842, Sir Richard Owen (1804–1892) assumed dinosaurs all descended from a common ancestor (in paleontological terms, they were a monophyletic group). This meant all dinosaurs shared some common characteristics. And when finally determined, the characteristics could be used to differentiate true dinosaurs from other organisms.

The English zoologist Richard Owen coined the term "dinosaur" back in 1842 (iStock).

In 1887, Harry Seeley (1839–1909) discovered there were two major dinosaur groups: the saurischians and the ornithischians. This led scientists to assume there was more than one common ancestor. To further complicate matters, discoveries of diverse dinosaur species pointed to many ancestors. As a result of this, dinosaurs were seen as a group of reptiles with few characteristics in common. Because of these earlier scientific interpretations, dinosaurs no longer shared one set of common characteristics; thus, they were classified as a polyphyletic group, arising from many sources among the archosaurs. They became a polyphyletic group, arising from many sources among the archosaurs. Interpreted this way, there was no longer one set of common characteristics that dinosaurs shared.

How many **species of dinosaurs** are currently known?

Not all dinosaur species that have been found have been named. Currently, there are between 600 and 700 known dinosaur species. But there is a major caveat: only about half of these specimens are complete skeletons and usually only complete (or nearly complete) skeletons allow scientists to confidently say the bones represent unique and separate species. Amazingly, many scientists speculate there may be between 700 and 900 more dinosaur genera that have yet to be discovered.

How many **species of dinosaurs** have been **named** since the first skeletons were uncovered?

The number of species named depends on the text—from 250 to more than 1,000 have been mentioned. Suffice it to say that hundreds of dinosaurs have been named since the first skeletons were dug up in the nineteenth century. Between the lack of consensus in terms of species and new fossils found every year, unraveling dinosaur patterns of evolution has been a major obstacle in dinosaur research.

How do the number of known **dinosaur species compare** to some modern species?

Even if there are truly 700 valid dinosaur species, the number is still less than one-tenth the number of currently known bird species; less than one-fifth the number of known mammal species; and less than one-third the number of known spider species.

How is a **dinosaur** currently **defined** using **cladistic analysis**?

Cladistics is a method of classifying all organisms by a common ancestry, and it is based on the branching of the organism's evolutionary family tree. Those that share common ancestors—and thus have similar features—fall into taxonomic groups called clades. Thanks to cladistic analysis, all dinosaurs were found to have many unique characteristics in common. In fact, these reptiles are defined as a monophyletic group descending from a common ancestor. And with the development of modern cladistic "testing," true dinosaurs can be distinguished from their closely related, but non-dinosaur, contemporaries.

Using cladistic analysis, a reptile is a dinosaur if it has several specific characteristics in its fossilized skeleton, including some of the following: an elongated deltopectoral crest on the humerus; three or fewer phalanges in the fourth finger of the hand; the absence of a postfrontal bone; a crest on the tibia; three or more sacral vertebrae; a fully open hip socket; a ball-like head on the femur; and a well-developed ascending process on the astragalus, fitting on the front face of the tibia. In other words, identifying a dinosaur has a great deal to do with its skeletal anatomy. Although these characteristics are very technical, the main point is that scientists now have a clear test to determine if a fossil skeleton is truly that of a dinosaur.

How are **dinosaurs named**?

Dinosaur names come from a number of places, but in general, they are named after a characteristic body feature (for example, the *Hypsilophodon,* or high-crowned tooth); after the place in which the first bones were found (for example, the *Mut-*

taburrasaurus); or after the person(s) involved in the discovery (for example, the *Leaellynasaura*).

In most cases, the names include two Greek or Latin words, or even combinations of the words. For example, *Tyrannosaurus rex* is a combination of Greek and Latin translated as "king of the tyrant lizards." Overall, the two names, known as the genus and species names, are used by biologists to describe all organisms on Earth, such as humans (*Homo sapiens sapiens*), domestic dogs (*Canis familiaris*), or rattlesnakes (*Crotalus horridus*).

How did **dinosaurs evolve**?

It's hard to determine which charts are right or wrong when it comes to dinosaur evolution, mainly because, since 1980, over 150 evolutionary trees of dinosaurs have been published, most of them looking at small groups of species.

How are dinosaurs **classified**?

The following table explains how dinosaur species are organized by order, suborder, infraorder, and family.

Classifications for Dinosauria*

Order	Suborder	Infraorder	Families
Saurischians	Theropods	Herrerasauria	Saltopodidae; Staurikosaurids; Herrerasaurids
		Ceratosauria	Coelophysids; Ceratosaurids; Podokesaurids; Abelisaurids; Noasaurids; Segisauridae
		Coelurosauria	Coelurids; Dryptosaurids; Compsognathids; Oviraptors; Caenagnathids; Avimimids; Ornithomimids; Garudimimids; Deinocherids; Dromaeosaurids; Troodontids; Tyrannosaurids
		Carnosauria	Allosaurids; Carcharodontosaurids; Spinosaurids; Baryonychids; Megalosaurids
		Segnosauria	Therizinosaurids; Segnosaurids
	Sauropods	Prosauropoda	Anchisaurids; Plateosaurids; Melanorosaurids; Massospondylidae
		Sauropoda	Cetiosaurids; Camarasaurids; Dicraeosaurids; Euhelopodids; Titanosaurids; Diplodocids; Brachiosaurids
Ornithischians	Ornithopods	Fabrosauria,	Heterodontosaurids; Fabrosaurids

Order	Suborder	Infraorder	Families
		Lesothosauria	
		Ornithopoda	Hypsilophodontids; Dryosaurids; Iguanodontids; Camptosaurids; Hadrosaurids; Lambeosaurids; Thescelosaurids
	Marginocephalia	Pachycephalosauria	Pachycephalosaurids; Homalocephalids
		Ceratopsia	Protoceratopsids; Ceratopsidae; Psittacosaurids
	Thyreophoda		Scutellosaurids; Scelidosaurids
		Stegosauria	Huayangosaurids; Stegosaurinae; Stegosauridae
		Ankylosauria	Nodosaurids; Ankylosaurids

*This is only *one* representation of how dinosaurs may be classified. There are actually many other dinosaur classifications, depending on the scientific study, and there will continue to be more as fossils are discovered and interpreted.

Is there a continent where **dinosaur** fossils have **never been found**?

No. It was once thought that Antarctica was the only continent that did not have any dinosaur fossils. But in December 2003, researchers working in separate sites thousands of miles apart in Antarctica found what they believe are the fossilized remains of two species of dinosaurs previously unknown to science—one a primitive sauropod. In 2007, yet another dinosaur discovery was made in Antarctica: a new genus and species of dinosaur from the early Jurassic—a massive, plant-eating, primitive sauropodomorph called *Glacialisaurus hammeri* that lived about 190 million years ago.

What **kinds of dinosaurs** were in the **saurischian** group?

The saurischians were a diverse group of dinosaurs, with both carnivores and herbivores. They exhibited two-legged and four-legged means of propulsion. The carnivores included the large two-legged *Allosaurus, Ceratosaurus, Tarbosaurus,* and *Tyrannosaurus*. There were smaller, two-legged carnivores, too, such as *Ornithomimus* and the *Dromaeosaurus,* which had specialized feet and their unique, slashing, raptorial claws. The herbivorous saurischians that are best known are the large, four-legged sauropods, the largest dinosaurs to have walked Earth. These dinosaurs had long necks and tails, with relatively tiny heads. Included in this group are *Brachiosaurus, Camarasaurus, Diplodocus, Mamenchisaurus,* and *Seismosaurus*.

What **kinds of dinosaurs** were in the **ornithischian** group?

The ornithischians were all herbivores and had two- and four-legged types. There were the four-legged armored dinosaurs such as *Ankylosaurus* and *Stegosaurus.*

There were large, horned dinosaurs such as *Eucentrosaurus* and *Triceratops*. Two-legged types included *Iguanodon*, and many of the duck-billed dinosaurs (hadrosaurs), such as *Corythosaurus*, *Lambeosaurus*, and *Maiasaura*.

Swedish naturalist Carolus Linnaeus devised the classification system for plants and animals that is still used today, both for living species and for species of the past such as dinosaurs (iStock).

What **events** led to the **dominance** of dinosaurs in the **Mesozoic** era?

Approximately 250 million years ago, at the end of the Permian period, or the beginning of the Triassic period (and thus, the end of the Paleozoic era and the beginning of the Mesozoic era), there was a mass extinction. This extinction eliminated close to 90 percent of all the species present on our planet (extraordinarily close to total extinction on Earth). The extinction was not selective; it eliminated organisms in the oceans and on land, including many invertebrates, armored fish, and reptiles.

The true reasons for the extinction are unknown, although there are several theories. One is that a collision with an asteroid or comet caused dust and debris to fly into the upper atmosphere, cutting off sunlight and radically changing the global climate. Another idea is that the moving continents changed the climate, sea levels, and thus, habitats, causing some species to change and adapt, while others died out. Still another theory focuses on Siberian flood basalts, in which tons of volcanic material erupted over a huge area in Asia toward the end of the Permian period, changing the climate and certain habitats.

Whatever the scenario, species that survived did so by adapting to the ecological niches that became vacant, allowing them to further evolve. After the Permian extinction, and throughout the Mesozoic era, it was reptiles in general, and the dinosaurs specifically, that diversified the most and became the dominate species on the planet.

How **long** were dinosaurs **dominant** on Earth?

The dinosaurs were the dominate species on Earth for approximately 160 million years. The Mesozoic era, often referred to as the "age of the reptiles," lasted from approximately 250 to 65 million years ago. It includes the Triassic, Jurassic, and Cretaceous periods.

Were there **dinosaurs** at the **beginning of the Mesozoic**?

It is interesting to note that at the beginning of the Mesozoic, there were no true dinosaurs; other reptiles dominated the landscape. But by the end of the Triassic

Within what percent of the geologic time scale did dinosaurs live?

Scientists estimate that dinosaurs existed on Earth for about 160 million years. Thus, dinosaurs were only around 3.1 percent of the time that has passed since the formation of Earth.

period, the dinosaurs became dominant, and they stayed that way for around 160 million years. Dinosaurs were not the only form of life that existed during this time. For example, there were smaller, lizard-like reptiles, small early mammals, insects, amphibians, invertebrates, and a wide variety of plants. In fact, these organisms helped the dinosaurs to stay in charge because many of the dinosaurs used this abundance of life for their sustenance and growth.

When did the **age of dinosaurs end**?

The age of the dinosaurs came to an end approximately 65 million years ago. From this point in time, there are currently no known dinosaur fossils. The time of the great dinosaur (and other species) extinction is used by scientists to delineate the end of the Cretaceous period as well as the end of the Mesozoic era. After this point, the Cenozoic era begins, starting with the Tertiary period.

TRIASSIC PERIOD

What is the **Triassic period** and how did it **get its name**?

The Triassic period follows the Permian period on the geological time scale. During this time, dinosaurs first began to evolve from the early reptiles, the first primitive mammals appeared, and the armored amphibians and mammal-like reptiles died out. The Triassic was one of the first labeled divisions on the geologic time scale, and it is the first of three periods (the others are the Jurassic and Cretaceous) making up the Mesozoic era.

The Triassic was first named in 1834 by German geologist Friedrich August von Alberti (1795–1878) to describe a three-part division of rock types in Germany. It was originally called the Trias, and is still called this by many European geologists. It is named after three, or "tri," layers of sedimentary rocks representative of the time period: from bottom to top, a sandstone, limestone, and copper-bearing shale. The three distinct rock formations are, from the bottom up, the Bunter (mostly Early Triassic), the Muschelkalk (Middle Triassic), and the Keuper (mostly Late Triassic).

How long did the **Triassic period last**?

The geologic time scale is not exact, and depending on the country or scientist, the dates of the Triassic period can vary by about 5 to 10 million years. On the average, the Triassic period is said to have lasted from about 250 to 205 million years, for a total of about 45 million years in length.

What are the **divisions** of the **Triassic period**?

In general, the informal way to define parts of the Triassic period is to use the terms early, middle, and late Triassic. More formally, they are capitalized (Early, Middle, and Late), or sometimes (Lower, Middle, and Upper) and include subdivisions within those groupings. The following lists a general interpretation of the Triassic epochs (although note that many researchers use slightly different notations; for example, many do not list the Rhaetian Age):

Triassic Period

Epoch	Age	Millions of Years Ago (approximate)
Early	Olenekian	245–242
	Induan	250–245
Middle	Ladinian	234–227
	Anisian	242–234
Late	Rhaetian	210–205
	Norian	221–210
	Carnian	227–221

What did the **Triassic period signify**?

The Triassic period represented the time after the great Permian period extinctions. It also was important as a time of transition—when the old life of the Paleozoic era gave way to the more highly developed and varied form of life of the Mesozoic era. The Permian period extinctions wiped out most of the animals and plants on Earth (about 90 percent of all species), making the very early Triassic an eerie place, almost completely devoid of the abundant life that existed perhaps hundreds or thousands of years before. Certain flora and fauna still dotted the land, and eventually, after about 10 million years or more, life began to emerge in full force again. But it still took even longer for larger animals, coral reefs, and other specialized animals to recover or evolve after the extinction at the end of the Permian period.

CONTINENTS DURING THE TRIASSIC PERIOD

Do Earth's **continents change positions**?

Yes, the continents continually change positions, but it takes them millions of years to shift and move great distances. Earth's continents are actually part of the thick plates that make up the planet's crust, all of various sizes and shapes. These plates fit together like a jigsaw puzzle. They do not move fast—only fractions of an inch to inches per year.

What are **continental drift** and **plate tectonics**?

The reason (or reasons) for Earth's crustal movement is still somewhat of a mystery. The most accepted theory of plate movement is called continental drift, and the theory of its mechanism, plate tectonics. These theories suggest that the continental plates move laterally across the face of the planet, driven by the lower, more fluid mantle. At certain plate boundaries, molten rock from the mantle rises at a mid-ocean ridge (such as the Mid-Atlantic Ridge, a long chain of volcanic mountains that lie under the Atlantic Ocean); or its equivalent on land, the rift valley (such as the one in eastern Africa), the magma solidifying and moving away to either side of the ridge. At other plate boundaries, plates are pushed under an adjacent plate, forming a subduction zone, in which the crust sinks into the mantle again. And at other boundaries, plates just slip by each other, such as the San Andreas Fault in California, in which a part of the North American plate slides by the Pacific plate.

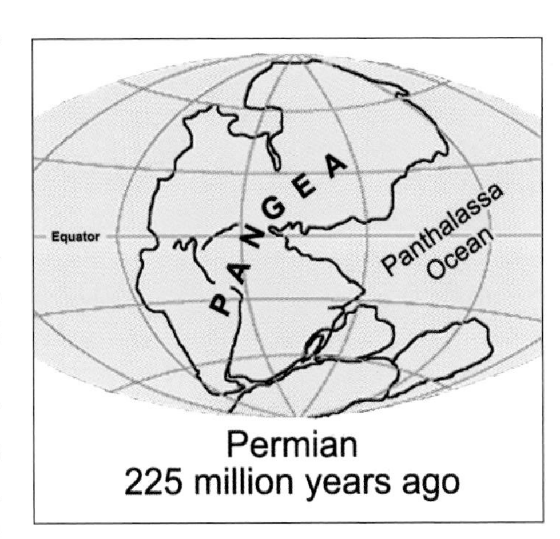

Hundreds of millions of years ago the seven continents were joined together as a supercontinent scientists now call Pangea. It later broke up into Laurasia and Gondwanaland during the Triassic, and eventually split apart even more (based on a U.S. Geological Survey map).

But not everyone agrees on these theories. One reason is because, although the idea of moving plates seems sound, the mechanisms for developing plate tectonics is not fully understood. Therefore, some scientists believe in continental drift, but not plate tectonics. Many of these scientists believe that the reason that the continents shift is that Earth is actually expanding, causing a false illusion of movement (although no one can explain why or how Earth is expanding). Another hypothesis is called "surge tectonics," in which the features of Earth's surface are explained by the sudden surge of plate movement, as opposed to a constant flow by the steady movement of the mantle. And still others suggest that the continents have always been in the same positions.

No one really can fully explain the reason for the continual movement of the plates. One thing is certain: the plates do move. Since the advent of Earth-orbiting observation satellites, scientists have been able to track the plates using sophisticated laser-ranging instruments that measure the minute movements.

How did **fossil evidence** support the theory of **continental drift**?

Scientists discovered the fossils of many identical-appearing species on widely separated continents. They had two theories for this. First, they theorized that separate species developed identically across the far-flung continents, a notion that was highly unlikely. The second theory was that the continents had been in contact with each other millions of years ago, and had somehow drifted apart.

For example, fossils found in South America were related to those in Australia and Antarctica. These landmasses were in contact sometime in the past, allowing species to roam freely, die, be buried, and become fossilized across these continents. The fossilized remains in the rock layers of the continents drifted with the landmasses, leading to widely separated—but nearly identical—fossils.

When did scientists determine the jigsaw puzzle fit of the continents?

The actual connection between the continental fit (the idea that continents fit together like the pieces of a jigsaw puzzle) was first proposed in 1858 by Antonio Snider-Pellegrini (1802–1885). Other scientists mentioned this idea for years afterward, but it was not until 1912 that German meteorologist and geologist Alfred Wegener (1880–1930) expanded the theory, suggesting that the continents at one time formed a supercontinent he called Pangea (or Pangaea). Wegener's theory was not taken seriously until about the 1960s, when scientists believed they had finally worked out a mechanism (plate tectonics) for the movement of the continental plates.

Who discovered **seafloor spreading**?

Harry Hess (1906–1969), an American geologist and professor of geology at Princeton University, discovered seafloor spreading. Based on material brought up from the ocean floor during a drilling project, he determined that rocks on the ocean floor were younger than those on the continental landmasses. He also discovered rocks on the ocean floor varied in age: there were older rocks farther from the mid-ocean ridges and younger rocks around mid-ocean ridges. Hess proposed that the seafloor was spreading as magma erupted from Earth's interior along the ocean's mid-ocean ridges. The newly created seafloor slowly spreads away from the ridges, and later sinks back into Earth's interior around deep-sea trenches.

What **magnetic evidence** did scientists use to verify **seafloor spreading**?

When molten lava is expelled from mid-ocean ridges, it cools, creating new ocean floor. And as the rock cools, specific minerals with magnetic properties line up with the prevailing magnetic field of Earth. This preserves a record of the magnetic field orientation at that particular point in time. Changes in the rocks' magnetic field records, called magnetic anomalies, happen when Earth's field reverses—or when the northern and southern magnetic poles change places—usually over hundreds of thousands of years. Scientists still do not know what causes these magnetic reversals, but it may have something to do with the giant convection currents in Earth's interior.

The theory of seafloor spreading was confirmed by measuring such magnetic anomalies in rocks on the ocean floor. Scientists discovered a symmetrical, striped pattern of magnetic anomalies on the ocean floor, spreading out on either side of the Mid-Atlantic Ridge. This ridge is a long volcanic mountain range that runs down the Atlantic Ocean seafloor between the continents of North America, Europe, Africa, and South America. The pattern and distribution of these stripes showed that the magnetic fields had reversed many times over millions of years—

and only could have formed if the seafloor had been spreading apart over those millions of years.

What is the driving **force** behind the movement of the continental plates?

Not everyone agrees on why the continental plates move across Earth—but there are some theories. In general, the continental plates are made of light material that "floats" on the heavier, molten material of Earth's interior (called the mantle). As the upper part of the mantle circulates and moves, it slowly "carries" the plates around the planet.

Continental plate movements have a variety of effects upon the planet, ranging from shifts in temperature to devastating earthquakes. Plate tectonics affect us today, just as they affected the dinosaurs millennia ago (iStock).

What are the **consequences** of **continental plate movements**?

As the term implies, the movement of the continental plates changes the positions of the continents. Along the continental boundaries, volcanoes and mountains form as the plates interact with each other. Some continents slowly crash into one another, forming huge mountain chains, such as the Himalayas in Asia (from the collision of the Indian and Asian plates). Other plates slide under one another in areas called subduction zones. The Andes Mountains are the result of a subduction zone between the Nazca and South American plates. Still other plates slip right by one another, such as the Pacific and North American plates. In this case, the slipping of the plates creates the San Andreas fault in California.

But there are other consequences of continental plate movement. In particular, this process also opens and closes the seas, changing ocean currents—and thus climates—around the world. In addition, volcanoes can form as plates sink under each other, and earthquakes can occur.

What did our **planet look like** at the start of the **Triassic period**?

Similar to today, most of the planet during the Triassic period was covered by ocean, but the distribution of the landmasses was not the same. Scientists believe there was essentially one large expanse of water called the Panthalassa Ocean. It surrounded the one very large landmass, or supercontinent, called Pangea, meaning "all Earth." This giant landmass straddled the planet's equator roughly in the form of a "C"; the smaller body of water enclosed by the "C" on the east was known as the Tethys Sea (or Tethys Ocean). Only a few scattered bits of continental crust were not attached to Pangea, and lay to the east of the larger continent. They included pieces of what we now call Manchuria (northern China), eastern China, Indochina, and bits of central Asia. In addition, the sea level was low, and there was no ice at the polar regions.

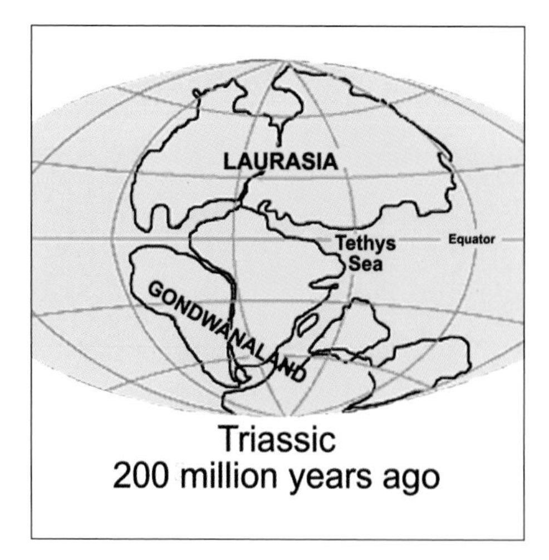

As Pangea began to break up, it formed two smaller supercontinents that scientists have named Laurasia and Gondwanaland (based on a map from the U.S. Geological Survey).

What led to the original **formation of the supercontinent Pangea**?

The same process that would eventually break apart Pangea led to its formation—the continents seemingly moving around the planet like icebergs on an ocean. There were two large landmasses on Earth during the Paleozoic era—Laurasia (North America and Eurasia) to the north of the equator, and Gondwanaland (or Gondwana, including South America, Africa, India, Antarctica, and Australia) to the south of the equator. These two continents slowly collided during the Late Paleozoic era, forming the supercontinent of Pangea. By the beginning of the Mesozoic era, Pangea was still the only true continent on the planet.

How did the **supercontinent Pangea change** during the **Triassic period**?

In the Early Triassic period, Pangea gradually began to break apart into two major continents again, the result of a seafloor-spreading rift. (This rift was similar to today's Mid-Ocean Ridge in the Atlantic Ocean, a volcanic seam that continues to spread, and along which the volcanic island of Iceland was born.) The Triassic rift extended westward from the Tethys Sea across what is today the Mediterranean Sea. The action of this rift separated northern Laurasia from southern Gondwanaland, which would eventually lead to the opening of the proto-Atlantic (or early Atlantic) Ocean. As North Africa split from southern Europe, there was a gradual rise in sea level that flooded south and central Europe.

Towards the Middle and Late Triassic periods, the spreading rift between North Africa and Europe grew westward, and it began to separate North Africa from the eastern part of North America. The resulting rift valley was the first true stage in the formation of the proto-Atlantic Ocean.

How did this **continental configuration** affect the **dinosaurs**?

During the Early Triassic period, the supercontinent Pangea allowed the precursors of dinosaurs to roam all over the huge landmass. As the supercontinent slowly split in two, it eventually cut off certain emerging dinosaur species from other species. The movements also caused certain ocean areas to widen and led to the inundation of parts of the landmasses, thus changing shorelines. This changed the types of vegetation and animal life in certain regions.

It is difficult to mention all the minute changes that occurred over such a broad expanse of time. But there are generalities: for example, as the Atlantic Ocean opened, lakes that eventually became the ocean grew larger and smaller—and even divided. These changes were often accompanied by an absence or abundance of sea life. Shorelines often grew rich in plant and animal life. We know that some dinosaurs came down to eat and drink around these changing lakes, as their tracks have been discovered in sediment that once lay along the shorelines.

Where are **Triassic period rocks** found?

Layers of Triassic period rocks are found in many countries around the world. They occur in certain localities in eastern and western North America, South America, the British Isles, western Europe, Asia, Africa, and Australia. The thickest Triassic period rock layer so far discovered lies in the Alps, measuring about 25,000 feet (7,500 meters) thick.

What is the **Newark Supergroup** and why is it **significant**?

The Newark Supergroup is a layer of Triassic period rocks located in the eastern United States; it is famous for its rocks and fossils from this period. The rock layers represent the remnants of several thousands of feet of sedimentary and volcanic rocks deposited in a chain of basins over a span of 45 million years. This layer is found in many locations, including New Jersey, Virginia, and North Carolina. The sedimentary strata contains a good cross-section of fossils from the Late Triassic period, including insects, fish, turtles, archosaurian reptiles (including dinosaurs, lizards, and snakes), lissamphibians (frogs, salamanders, and caecilians), and numerous plant fossils. Paleontologists hope to find additional vertebrate fossils in the Newark Super-

One of the richest fossil deposits from the Triassic period, the Newark Supergroup is located in the eastern United States (iStock).

group layers that will shed more light on the evolutionary stages that led to groups of many organisms—especially the dinosaurs—during the Triassic period.

The Newark Supergroup—indeed most of the area of the rift valley called the Atlantic Rifting Zone—is famous for reptile footprints, with tens of thousands of extremely well-preserved tracks from the Triassic and Jurassic periods. The tracks were first recognized in 1836 by U.S. paleontologist, the reverend Edward Hitchcock, a professor of natural theology and geology at Amherst College. He also described the Connecticut River Valley dinosaur footprints. These tracks are important because they show evidence of the evolution and eventual dominance of the dinosaurs.

What is the **Ischigualasto formation** and why is it **important**?

The Triassic rocks of the Ischigualasto formation of Argentina are thought to be some of the richest fossil deposits, and include the earliest known, well-preserved fossils of such dinosaurs as *Herrerasaurus ischigualastensis, Eoraptor lunensis,* and *Pisanosaurus mertii*. The drab, gray rocks of this formation were deposited in a humid environment. The area where this formation is found—the Ischigualasto Valley, or the "Valley of the Moon"—is located to the east of a subduction zone. Consequently, it also received deposits of volcanic ash during the Triassic. Dating of this ash has enabled scientists to determine an age of approximately 228 million years for this formation, placing it in the Middle to Late Triassic periods.

Vertebrate land fossils are abundant in the formation, including skeletons of advanced forms of synapsids, and odd beaked reptiles called rhynchosaurs. There are also large carnivores called rauisuchians, and smaller archosaurs, but the most prevalent fossils in the formation are of dinosaurs. In fact, based on these dinosaur fossils and ones found in the Santa Maria formation of Brazil, some scientists suggest that perhaps dinosaurs arose in South America.

TRIASSIC DINOSAURS

Did any **Triassic period dinosaurs fly**?

No. There were flying and gliding reptiles throughout the entire Mesozoic era—but never any flying dinosaurs.

Did any **Triassic period dinosaurs** live in the **oceans**?

No. There were a variety of marine reptiles in the water, but at no time did marine dinosaurs exist. The entire classification of dinosaurs is limited to land-dwelling reptiles with specific characteristics.

Were there **many dinosaurs** during the **Triassic period**?

Based on the current findings in fossil records, there were not many dinosaurs living during the Triassic period. Dinosaurs began to evolve from the reptiles toward

What were the first true dinosaurs like?

Based on current fossil finds, the first true dinosaurs emerged in the Late Triassic period, between about 230 to 225 million years ago. They were small, agile, carnivorous reptiles whose unique characteristics, such as two-legged motion, enabled them to quickly dominate the available ecological niches.

Apparently, by the time dinosaurs evolved, they had already split, even at this early stage, into two major groups: the ornithischians and the saurischians. These two main groups of dinosaurs are based historically on their hip-bone structure.

the end of the Triassic period, but great numbers of the creatures did not flourish until the Jurassic period.

Why did the **dinosaurs** begin to **thrive** in the **Late Triassic period**?

Scientists theorize that there were a number of reasons for the emergence of dinosaurs in the Late Triassic period. One reason was that dinosaurs evolved to become biologically superior. For example, they developed an erect posture, with bipedal (two-footed) motion. This gave them a longer stride and quickness, enabling them to catch and devour other semi-erect reptiles. Another adaptation might have been warm-bloodedness—although this idea is still controversial. If it is true that the reptiles developed a form of warm-bloodedness, it would have allowed them to be more active than their cold-blooded relatives.

Still other scientists suggest that these adaptations were not the reason for the dominance of dinosaurs. Instead, they believe a major extinction of certain therapsids (reptilian ancestors of mammals), rhynchosaurs (lizard-like reptiles), and early archosaurs (reptiles comprising dinosaurs, pterosaurs, and crocodilians) in the middle of the Triassic period opened up ecological niches that the dinosaurs then filled.

What were some of the **early dinosaurs**?

There were not many early dinosaurs in the Triassic period. Some of the following dinosaurs are represented by only a few fossil discoveries. This list will definitely change in the future as more dinosaur fossils are found and dated.

Examples of Early Triassic Period Dinosaurs

Name	Common Name	Approximate Age (million years ago)	Locality	Maximum Body Length (in feet/meters)
Anchisaurus	Near Lizard	200–190	USA	6.5/2
Coelophysis	Hollow Form	225–220	USA	10/3
Eoraptor	Dawn Hunter	225	Argentina	3/1

Name	Common Name	Approximate Age (million years ago)	Locality	Maximum Body Length (in feet/meters)
Herrerasaurus	Herrera Lizard	230–225	Argentina	10/3
Plateosaurus	Flat Lizard	about 210	France, Germany, and Switzerland	23/7

Why is ***Herrerasaurus ischigualastensis*** thought to be one of the **earliest dinosaurs** known?

Based on the characteristics of its skeleton—a blend of archosaur and dinosaur features—scientists believe the *Herrerasaurus ischigualastensis* was an early dinosaur. Initially, scientists knew this 10- to 18-foot (3- to 6-meter) carnivorous reptile had a skull and foot similar to other archosaurs such as the *Euparkeria*. Based only on this evidence, it was impossible to determine if the *Herrerasaurus* was truly a dinosaur. But using the modern cladistic test, the *Herrerasaurus* skeleton shows all but a few of the characteristics that define dinosaurs. In particular, *Herrerasaurus* lacks bone in the center of the hip socket, a key characteristic that defines this reptile as a true dinosaur—and one of the earliest known.

Is the ***Eoraptor lunensis*** also one of the most **primitive dinosaurs**?

Yes, as of this writing the *Eoraptor lunensis* ("dawn hunter") fossils discovered in the Ischigualasto formation in Argentina, South America, are thought to be those of one of the most primitive dinosaurs known (the other is the *Herrerasaurus ischigualastensis* described above). This is because the *Eoraptor* fossils are a mix of primitive and specialized characteristics—a blend of features expected for the "first" dinosaur. The *Eoraptor* had a three-fingered hand, which connected them to the theropod dinosaurs. But *Eoraptor* fossils lack the specialized features that would place this reptile in one specific major group of dinosaurs.

Which **Triassic period dinosaurs** were **herbivores**?

There were numerous herbivore dinosaurs that evolved at the end of the Triassic period. Fossils of the *Thecodontosaurus* ("socket-tooth lizard"), a small herbivore, was one of the first Triassic dinosaurs ever found; it was reported in England in 1836. The *Plateosaurus* ("flat lizard") was also a plant-eating dinosaur that lived during the Triassic period; the first fossils were found in Germany in 1834, but it took three more years before they were described. The creatures were considered one of the largest known dinosaurs of their time, and had peg-like teeth and huge thumb claws that were perhaps used to gather plants from taller trees.

What is the **earliest-known herbivorous dinosaur**?

The earliest-known, plant-eating dinosaur was the *Pisanosaurus mertii,* found in the Ischigualasto formation of Argentina, South America. The dinosaur is dated at

One of the first predatory dinosaurs, *Coelophysis* was the first animal to show herding behavior, according to some paleontologists (iStock).

approximately 230 million years old (Late Triassic). And like its relatives *Herrerasaurus ischigualastensis* and *Eoraptor lunensis,* the *Pisanosaurus* was relatively small (around 3 feet [1 meter] in length), lightweight, and bipedal.

Which **Triassic period dinosaurs** were **carnivores**?

The first carnivorous dinosaurs in the Triassic period were the *Eoraptor* and *Herrerasaurus*. Both dinosaurs were small and bipedal, with powerful hind limbs and long tails for balance. The later *Coelophysis* was also a meat-eater. Some of the first fossils of this dinosaur were found in the southwestern United States—and although it is debated, some scientists believe the creatures were the first to show evidence of a herding behavior.

What was ***Coelophysis*** like?

Coelophysis ("hollow form") appeared around 215 million years ago. It was a small, relatively delicate dinosaur, measuring up to 10 feet (3 meters) in length, with an elongated neck, strong grasping hands, and a long, slender skull with sharp teeth. As it walked on two legs, *Coelophysis* used its long, slender tail for balance. Its feet were narrow, with only three prominent toes, a distinctive characteristic of the theropods. The *Coelophysis* were part of a carnivorous dinosaur group that later included the *Tyrannosaurus rex* and *Velociraptor*.

What were the **smallest dinosaurs** known to have existed during the **Triassic period**?

At present, the smallest Triassic period dinosaurs known to have existed were the early, carnivorous bipeds *Eoraptor* and *Herrerasaurus*. The *Eoraptor* was about 3 feet (1 meter) long, whereas the *Herrerasaurus* was about 10 to 18 feet (3 to 6 meters) in length.

Were there other reptiles in the **Middle to Late Triassic** period that may have been **early dinosaurs**?

Yes, the *Herrerasaurus* may have had some other dinosaur relatives in North and South America during the Middle to Late Triassic periods. Some scientists believe the *Staurikosaurus pricei* from southern Brazil and northwestern Argentina, and the *Chindesaurus bryansmalli* from the Chinle formation in North America, were close relatives. But disagreements still exists as to where to put these dinosaurs on the overall family tree. One thing is certain: this important group gives us some idea of the time that dinosaurs first appeared—and what these early creatures looked like.

What was the **largest dinosaur** known to have existed during the **Triassic period**?

At present, the largest dinosaur known to have existed during the Triassic was the herbivorous *Plateosaurus,* with a length of up to 20 to 33 feet (6 to 10 meters). The creature had a long neck, large stocky body, and a pear-shaped trunk. Its skull was deeper than that of *Coelophysis,* though still small and narrow compared to the size of its body. The reptile's teeth were set in sockets, and were small, peg-like, and leaf-shaped, with coarse serrations. The eyes were directed to the sides, rather than to the front, which reduced its depth perception, but gave the *Plateosaurus* a wider field of view to detect predators.

The foot of the *Plateosaurus* was very similar to that of *Herrerasaurus* and the structure of the legs indicates that this dinosaur was not a fast runner. A unique feature of *Plateosaurus* was its broad, apron-shaped pubis, forming a "shelf" that may have provided support for its huge gut. The animal's tail could be bent sharply upward near its base—a good characteristic to have when rearing up against trees.

OTHER LIFE DURING THE TRIASSIC

If the dinosaurs were just evolving, which **land and marine animals** dominated the **Triassic period**?

On land, the true dominant species of the Triassic period, even after the dinosaurs started to evolve during the Late Triassic, were the non-dinosaurian predators, the archosaurs; the main herbivores were the dicynodont (synapsids). In the oceans, many types of reptiles and fishes dominated.

What are the **tetrapods**?

Tetrapods ("four feet") is a term used to describe the four-legged creatures that left the water to live on land. The first tetrapods were the amphibians; dinosaurs were also tetrapods. In fact, all modern amphibians, reptiles, birds, and mammals are tetrapods. (Interestingly, some scientists say humans can be considered tetrapods, as they are descendants of the original tetrapods; we have four limbs, though we use two of them as arms instead of for walking.)

All the approximately 5,000 species of frogs known today are skilled jumpers, and they are all apparently descended from *Triadobatrachus,* which lived during the Triassic (iStock).

Where did **tetrapods live** on **Pangea** during the **Late Triassic period**?

The distribution of the tetrapods was not uniform during the late Triassic period—even though Pangea was a huge connected landmass. The major reason for this non-uniform distribution was the very pronounced climate zones over the very large continental landmass. The equator had a narrow, humid zone; humid temperate climates existed from around 50 degrees north and south latitudes to the poles. In between, at approximately 30 degrees north and south latitude, were wide arid zones.

The distribution of tetrapods followed these climatic variations. For example, the prosauropods, a group of dinosaurs that included the *Plateosaurus,* had a range roughly corresponding to the temperate zones in both hemispheres; prosauropods shared this area with large amphibians. Phytosaurs, distant relatives of the crocodiles, were limited to the Northern Hemisphere and the coastal regions of the Southern Hemisphere. And some tetrapods, such as rauisuchians, crocodylomorphs, and aetosaurs, were distributed over all of Pangea.

What were the **major groups** of **land organisms** during the **Triassic**?

The true numbers, types, names, and evolutionary events of the Triassic land animals is often highly debated, which is typical when we try to interpret our ancient past. The following list gives a general synopsis of only some of the larger animals; however, not everyone agrees on these interpretations. Fossils are subject to various explanations that sometimes vary from scientist to scientist—often making it difficult to arrive at any definitive statements about these animals. Below are brief descriptions of the other animals living during this time:

Amphibians

Primitive amphibians: Only a few large, primitive amphibians (labyrinthodonts), survived into the Mesozoic era after the Permian period extinctions; they gradually declined in abundance and diversity during the Mesozoic; most of them were aquatic, the majority living in freshwater environments.

Primitive frogs and toads: First links to these modern amphibians (or lissamphibians) evolve during the Early Triassic; the oldest member of the frog group was the *Triadobatrachus,* the only known link between the true frogs with jumping motion and the primitive ancestors of frogs.

Land Reptiles (Anapsids, Diapsids, and Euryapsids)

First turtles: Of the several Paleozoic groups of anapsids, only turtles and procolophonids survived into the Mesozoic; the oldest subgroup, proganochelydians, were moderately large, but the animal could not pull its head inside its shell; disagreement exists as to whether turtles are truly diapsids, not anapsids.

Procolophonids: Lizard-like in their overall habits and shape; they probably ate insects, smaller animals, and some plant material; even though they looked like lizards, the true lizards did not appear until the Late Jurassic.

Rhynchosaurs: Short-lived diapsid reptiles of the group Archosauromorpha; they were herbivorous, walked on all fours, and had huge beaks that helped them bite off vegetation; they are so widespread during the Triassic that their fossils are often used to correlate deposits on different continents.

Tanystropheids: Very short-lived diapsid reptiles of the group Archosauromorpha; they lived near (and sometimes in) marine waters; they had an oddly-shaped body, with a tiny head on an extremely long neck, and a short, medium-sized body; the reason for such a long neck is unknown, but one theory suggests that it helped the tanystropheid stretch its neck low over the water in order to catch fish.

Archosaurs: Part of the diapsid reptile group Archosauromorpha, and the dominant tetrapods on the continents during most of the Mesozoic; the archosaurs ("ruling reptiles") were the precursors to dinosaurs; they are characterized by their better adaptation of legs, feet, and hips, giving them agility on land; others categorize the archosaurs by the openings in their skull; the earliest archosaurs were relatively large and carnivorous, and either lived on land, or led a semi-aquatic existence.

Aetosaurs: Heavily armored, herbivorous archosaurs.

Phytosaurs: lived during the Late Triassic only, and looked very much like modern crocodiles.

Crocodylomorphs: A group that includes crocodiles, alligators, caimans, and gavlals, known to exist from the Late Triassic to the present; not all survived to the present, including the fast-running saltoposuchians.

Rauischians: the creature's upright front and hind legs were under the trunk of the body, making them the dominant land predator during the Triassic.

Ornithosuchian: Relatively large (10 feet [3 meters]), land predators that may have walked on all fours, but ran fast only on its hind legs; they were the most dinosaur-like of the non-dinosaur archosaurs.

Small rodent-sized mammals similar to this mouse first appeared as early as the Triassic period (iStock).

Ornithodira: The Middle and Late Triassic group of archosaurs to which the dinosaurs belong; it also includes the pterosaurs, birds, and some early forms of creatures that appear to be closely related to dinosaurs and pterosaurs.

Aerial Reptiles (Diapsids)

Gliding reptiles: The three main Late Triassic gliding reptiles used either skin membranes on the wings and legs (such as the *Sharovipteryx*), scales (*Longisquama*), or fan-like wings (*Kuehneosaurus*)—all of which acted as an airfoil, allowing the reptiles to glide through the air; they probably did not flap their "wings" for powered flight.

Flying reptiles: The pterosaurs (they are also called pterodactyls, but that is only one subgroup of pterosaurs); the front legs (or arms) were modified into true wings by the elongation of the fourth finger, which supported a skin membrane stretching to the body; they probably flapped their wings occasionally for powered flight; they lived from the ocean shores and inland, eating fish, insects, and other small animals; they evolved during the Late Triassic period.

Mammals and Their Reptile-like Relatives (Synapsids)

Therapsids: More advanced synapsids; varied group of mammal-like reptiles that apparently evolved from the pelycosaurs, the earliest known mammal-like reptile that evolved in the Late Carboniferous period, about 290 million years ago, and went extinct in the Late Permian period; the biggest change was their ability to walk more efficiently with their limbs tucked beneath their body, whereas pelycosaurs walked with their limbs in a sprawled position; one group of therapsids gave rise to mammals, known from the Late Triassic to today.

Anomodonts: The most common subgroup were the dicynodonts, large, herbivorous, mammal-like reptiles; it includes the *Lystrosaurus,* a three- to six-foot-

(one- to two-meter-) long, pig-like animal that has been found as fossils in Australia, South Africa, India, China, and Antarctica, and hippopotamus-like *Kannemeyeria,* a 10-foot- (3-meter-) long animal with two big canine-like teeth in the upper jaw; they died out during the Late Triassic period.

The *Elasmosaurus* was one of the largest plesiosaur species to swim through Earth's oceans. Today, some people believe that the Loch Ness monster in Scotland is a type of plesiosaur that survived the extinction of the dinosaurs (Big Stock Photo).

Cynodonts: Carnivorous, mammal-like therapsid reptiles; they walked more upright, with limbs held more underneath their bodies; some were probably wolf-like animals, and some seem to have had whiskers, pointing to the possibility of having fur, and thus, may have been warm-blooded; they evolved during the Late Permian to the Middle Jurassic; at least one group of cynodonts evolved into mammals.

Therocephalians: Existed from the Late Permian to Middle Triassic period, these therapsid reptiles had their peak during the Late Paleozoic era; they were small to medium sized, walked on all fours, and ate insects or small animals.

True mammals: Small, about the size of a rat or mouse, with the largest about the size of a cat; they were probably nocturnal; they probably ate insects or small mammals, and at least one group ate plants; they evolved in the latter part of the Triassic, at the same time as the dinosaurs first appeared.

Triconodonts: Late Triassic to Late Cretaceous mammals; one of the oldest fossil mammals; three cusps of teeth in a straight row give them their name.

Haramyoids: Late Triassic to Middle Jurassic mammals; one of the oldest fossil mammals; their teeth had many cusps in at least two parallel rows.

Other Creatures

Insects: Very common; included the first species to undergo complete metamorphosis from larva through pupa to adult.

Spiders: Very common; spiders had been around for millions of years already, showing up in fossils dating back to the Cambrian period.

Earthworms: Very common; earthworms had been around for millions of years already, showing up in fossils dating back to the Cambrian period.

What were the major **marine animals** living during the **Triassic period**?

There were many marine animals that swam the oceans of the Triassic, and many of the species still continue to this day. In general, the oceans held the following animals, although with new fossil discoveries, this list may eventually change.

All modern sea urchins, such as this red sea urchin off the coast of British Columbia, are descended from urchins that survived the Permian extinction (iStock).

Major Marine Animals

Reptiles (Euryapsids)

Ichthyosaurs: Also called "fish reptiles," these were predatory sea reptiles that probably preyed mostly on shellfish, fish, and other marine reptiles; they looked similar to, and probably had some of the same habits of, modern dolphins, whales, and sharks; they lived in the oceans from the Early Triassic to Middle Cretaceous periods, probably outcompeted by the mosasaurs of the Middle Cretaceous period.

Plesiosaurs: Medium to large, long- to short-necked reptiles, with bulbous bodies; their four legs were modified into paddles; they probably ate mostly fish; they lived mostly in marine environments, but some also lived in freshwater lakes; they lived from the Early Triassic to the end of the Cretaceous period and are often sighted as the model for what the Loch Ness monster is presumed to look like.

Placodonts: Large marine reptiles that had a long trunk and tail, with feet that were probably webbed; their teeth were using for crushing, and they probably ate clams and other shelled invertebrates from the ocean floor.

Nothosaurs: Small to moderate-sized marine reptile with long necks and sharp, conical teeth for spearing fish; their legs were modified flippers, rather than the paddle-shaped legs of the more advanced eurapsids; they lived from the Early to Late Triassic period.

Other Marine Creatures

Sea urchins: The few pencil urchins that survived the Permian period extinction are also the ancestors of all modern urchins; the Triassic period was also the time of the first burrowing urchins.

Corals: First relatives to the modern corals evolved during the Triassic period.

Crabs and lobsters (crustaceans): First close relatives of modern crabs and lobsters evolved during the Triassic period.

Ammonoids (chamber-shelled organisms): Ammonoids rapidly diversified during the Triassic period.

Bony fishes: Found in salt, brackish, and fresh water, and could often move back and forth among the three; they are divided into two groups based on their structure: the ray-finned (for example, the Triassic period's *Perleidus*) and lobe-finned (for example, the Triassic period's *Diplurus*).

Sharks: During the Triassic, the intermediate form between primitive and modern sharks evolved; the earliest sharks evolved during the Paleozoic era, middle Devonian period, about 130 million years before; one of the modern survivors of this group is the Port Jackson shark.

What is the **Jurassic period** and how did it get its **name**?

The Jurassic period follows the Triassic on the geological time scale. Though the dinosaurs had their origins and approximately 25 million years of evolution in the Triassic period, it was not until the Jurassic that this group really blossomed. This was the time when the giant, herbivorous sauropods like *Apatosaurus* roamed the land; when plated dinosaurs like *Stegosaurus* first appeared; and when large carnivorous species like *Allosaurus* preyed on the other dinosaurs. It was also when *Archaeopteryx*—a creature that many paleontologists consider to be one of the first ancestors of birds—flew through the air.

The name Jurassic comes from the Jura mountain range, a chain of mountains that straddle the border between France and Switzerland. It was there that the first Jurassic period sedimentary rock and accompanying fossils were found. The Jurassic is the second of three periods (the first is the Triassic and the last is the Cretaceous) making up the Mesozoic era.

What are the **divisions** of the **Jurassic period**?

The Jurassic period is normally divided into three main divisions, or epochs: Early, Middle, and Late; more informally, the period is labeled with lower case letters, or the early, middle, and late Jurassic. In addition, scientists often use the terms Lower, Middle, and Upper to describe the divisions of the Jurassic.

Each of these main epochs is further broken up into subdivisions. To make things more confusing, the smaller ages often have different names (and often dates), depending on whether you are using European, North American, or Australian and New Zealand nomenclature. Presented here is a general list of North American Jurassic period divisions. These dates are not absolute, and may vary slightly from source to source.

Jurassic Period

Epoch	Age	(approximate) Millions of Years Ago
Early	Navajo	195–178
	Kayenta	202–195
Middle	Twin Creek	170–163
	Gypsum Springs	178–170
Late	Morrison	156–141
	Sundance	163–156

How long did the **Jurassic period last**?

The Jurassic period lasted from approximately 200 to 145 million years ago, a time period of approximately 55 million years. The exact dates are debated, of course, and there are some variations of the dates in the literature, but the time frame is close.

What **event occurred** at the division **between** the **Triassic and Jurassic** periods, and why was this **important to the dinosaurs**?

There was apparently a major extinction around 200 million years ago between the Triassic and Jurassic periods. This event (or events that lasted 10,000 years) led to the almost complete disappearance of many marine groups, such as some of the ammonoids; as well as the complete disappearance of some reptiles, including some types of archosaurs, phytosaurs, aetosaurs, and rauisuchians. Though many scientists speculate that this extinction was caused by an asteroid impact, the suspected resulting crater, Manicouagan in British Columbia, Canada, has been dated at 10 million years too early. There are thus heated debates as to the causes of this extinction event.

Some scientists think that this end-of-the-Triassic-period extinction event opened up more ecological niches (how a species fits into its environment) into which the dinosaurs dispersed, allowing them to flourish and become dominant. However, others feel that the dinosaurs were already on their way to dominance due to the major extinction event at the end of the Permian period. The real sequence of these events may never be known, but in any case, the dinosaurs did start to become dominant during the early Jurassic period.

There is another theory that tries to explain the extinction event between the Triassic and Jurassic periods: There were very large lava flows for approximately 600,000 years close to the division between the Triassic and Jurassic periods—one of the largest such events known to have occurred on our planet. The side effects of these flows, such as the emission of carbon dioxide and sulfur aerosols, may have contributed to the mass extinctions at this time by changing the atmosphere's composition and/or climate.

Another suggestion is that an impacting asteroid or comet may have actually caused the lava flows, or worked in conjunction with the flows, to create an even harsher environment. The resulting environmental changes—from climate to vegetation—could have led to the mass extinction event between the Triassic and Jurassic periods.

A wide diverstiy of Jurassic period fossils have been found among the sedimentary rocks of the Morrison formation in Colorado (iStock).

What is the **Morrison formation** and where is it found?

The Morrison formation is a layer of sedimentary rocks famous for the number and diversity of their Jurassic period dinosaur fossils. This formation, named for Morrison, Colorado, is found throughout a large region of western North America. The number of Jurassic dinosaur fossils found in Utah's Morrison formation include *Allosaurus, Apatosaurus, Barosaurus, Brachiosaurus, Camarasaurus, Camptosaurus, Ceratosaurus, Diplodocus, Dryosaurus, Dystrophaeus, Marshosaurus, Stegosaurus, Stokesosaurus,* and *Torvosaurus*.

What is the **Tendaguru formation**?

The Tendaguru rock formation formed during the late Jurassic period, with the best outcrops found in Tanzania in East Africa. It was discovered in the early 1900s, and is considered one of the greatest dinosaur graveyards in the world. Several expeditions to the area were carried out by German and British scientists early in the twentieth century—and a few in more recent years.

The Tendaguru is often compared to the Morrison formation because the overall fauna are similar in both dinosaur fossil-rich layers of rock. For instance, *Brachiosaurus* fossils found in both formations are strikingly similar. One reason for these similarities is that the continental landmasses remained close together during the Jurassic period, allowing the dinosaurs to remain widely distributed. But there are differences, too, including the absence of large theropod dinosaur fossils in the Tendaguru when compared to the Morrison formation.

How did **dinosaurs** become so **prolific** between the **Triassic** and **Jurassic** periods?

Scientists believe that the end of the Triassic was one of the busiest times in the history of land vertebrates. There were all types of animals, such as crocodiles, turtles,

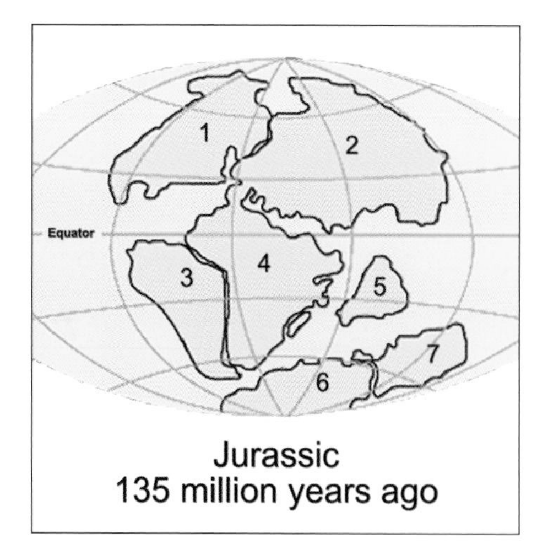

Laurasia and Gondwana began to break up during the Jurassic period. The numbers indicate what would become 1) North America, 2) Eurasia, 3) South America, 4) Africa, 5) the Indian subcontinent, 6) Antarctica, and 7) Australia.

lizard relatives, pterosaurs, therapsids, giant amphibians, the first mammals, and dinosaurs. But during a short period of time—maybe only about 5 to 10 million years—at the beginning of the Jurassic, dinosaurs began to dominate the land, filling almost every available environment.

There are several theories that attempt to explain this dinosaur dominance. The first one is competition, as the dinosaurs out-competed the other animals for food. Second is opportunism: the dinosaurs took advantage of their specialized characteristics to take over territories of other animals. In particular, scientists believe a major extinction of certain therapsids (reptilian ancestors of mammals), rhynchosaurs (lizard-like reptiles), and early archosaurs (reptiles comprising dinosaurs, pterosaurs, and crocodilians) in the middle of the Triassic period opened up ecological niches that the dinosaurs then filled.

Another suggestion is that the dinosaurs' specialized anatomy also allowed them to beat out competitors. For example, they developed an erect posture with bipedal (two-footed) motion, giving them a longer stride and quickness and enabling them to catch and devour the other semi-erect reptiles. The best examples of this were the *Tyrannosaurus rex* and other large carnivores. These creatures could walk upright because of the way their hips were put together, giving them an edge and allowing them to free their forearms to grasp prey; this was something few other animals could do.

Finally, another adaptation might have been warm-bloodedness, although this idea remains controversial. If it is true that the dinosaurs developed a form of warm-bloodedness, it would have allowed them to be more active than their cold-blooded cousins.

Where were the **continents** located during the **Jurassic period**?

In the Early Jurassic period the continents were still clustered around the equator roughly in the shape of a "C" that bordered the Tethys Sea. However, unlike the Triassic period, in which the continents were all part of one giant landmass known as Pangea, a split formed during the Jurassic period that divided Pangea into two large landmasses. The most accepted theory for the breakup of the supercontinent Pangea is the action of plate tectonics, causing a spreading rift split the "C" shape of Pangea into two large, separate landmasses.

What is the difference between Gondwanaland and Gondwana?

There really is no difference between the terms Gondwanaland and Gondwana; Gondwanaland was the original term used for the supercontinent. They are synonymous, and the use of the terms appears to be a personal preference.

What were **Laurasia** and **Gondwana**?

Laurasia and Gondwanaland, or Gondwana, were the two major continents (often called supercontinents) of the Jurassic period. As the gap between North America and Africa widened, driven by the spreading rift, so did the gap between North and South America. Water filled this gap, separating Pangea into the northern continent of Laurasia, and the southern continent of Gondwana. Despite the separation of the huge continent, scientists have found fossil skeletons of the *Brachiosaurus* and plated stegosaurs in Africa and North America. This indicates that although the continents were separating, there were probably land bridges that popped up from time to time, allowing the species to spread to both continents.

Laurasia included the present-day areas of Europe, North America, and Siberia. Also included in this large landmass was Greenland. Gondwana included the present day areas of Africa, South America, India, Antarctica, and Australia.

How did the northern continent of **Laurasia** and the southern continent of **Gondwana change** during the Jurassic period?

As the rift grew westward, separating Pangea into Laurasia and Gondwana, and millions of years passed, there were numerous changes on each of these large continents. Africa began to separate from Europe, starting the formation of the Mediterranean Sea. Italy, Greece, Turkey, and Iran were attached to the North African part of Gondwana, while Antarctica and Australia detached from Gondwana, but were still in contact with each other. And the block of land we now call India drifted northward.

North America separated from Gondwana and drifted west, resulting in the formation of the Gulf of Mexico, and the widening of the North Atlantic Ocean. South America and Africa began separating, creating a long, narrow seaway that would eventually become the South Atlantic Ocean. In addition, sea levels rose during this period, resulting in shallow seas flooding parts of North America and Europe in the late Jurassic.

Did **climate conditions** lead to **new dinosaur groups** in the **Jurassic** period?

Yes, scientists believe the climate turned milder in the Jurassic period and lush, tropical vegetation began to grow. This gave rise to new dinosaur groups, including the long-necked sauropods (plant-eaters). The increase in the amount of vegetation

With a new climate permitting, larger plants were able to grow. As a result, dinosaurs like the long-necked sauropods evolved to reach them (iStock).

allowed these animals to grow quite large. Their long necks allowed these animals to reach food that was out-of-touch for most other dinosaurs. With abundant food supplies, these plant-eating animals continued to grow from generation to generation.

IMPORTANT JURASSIC DINOSAURS

What are some of the major **dinosaurs** that lived during the **Jurassic period**?

There were two groups of dinosaurs during the Jurassic: the saurischians (reptile or lizard-hipped), divided into the sauropods (herbivores) and theropods (carnivores); and the ornithischians (bird-hipped), such as the stegosaurs, ankylosaurs, and ornithopods. This latter group were all herbivores. The following table lists some of the dinosaurs that lived during the Jurassic. New fossil finds are occurring all the time and the list will subsequently grow.

Jurassic Period Dinosaurs

Name	Common Name	Age (million of years ago)	Locality	Maximum Length (feet/meters)
Allosaurus	Other Lizard	150–135	USA	50/15
Apatosaurus	Deceptive Lizard	154–145	USA	70/21
Archaeopteryx*	Ancient Wing	147	Germany	1.5/0.5

Name	Common Name	Age (million of years ago)	Locality	Maximum Length (feet/meters)
Barosaurus	Heavy Lizard	155–145	USA	80/24
Brachiosaurus	Arm Lizard	155–140	USA/Tanzania	75/23
Camarasaurus	Chambered Lizard	155–145	USA	65/20
Camptosaurus	Bent Lizard	155–145	USA	16/5
Coelurus	Hollow Tail	155–145	USA	8/2.4
Compsognathus	Pretty Jaw	147	Germany	26 inches
Dacentrurus	Pointed Tail	157–152	France, England, Portugal	20/6
Diplodocus	Double Beam	155–145	USA	90/27
Dryosaurus	Oak Lizard	155–140	USA, Tanzania	13/4
Kentrosaurus	Spikey Lizard	140	Tanzania	10/3
Mamenchisaurus	Mamenchi Lizard	155–145	China	72/22
Massospondylus	Massive Vertebra	208–204	South Africa	13/4
Megalosaurus	Big Lizard	170–155	England, Tanzania	30/9
Ornitholestes	Bird Robber	155–145	USA	6.5/2
Scutellosaurus	Little Shield Lizard	208-200	USA	1/1.2
Patagosaurus	Patagonian Lizard	163–161	Argentina	60/18
Pelorosaurus	Monstrous Lizard	150	England	unknown
Scelidosaurus	Limb Lizard	203–194	England	13/4
Stegosaurus	Roof Lizard	155–145	USA	30/9
Tuojiangosaurus	Tuojiang Lizard	157–154	China	21/6.4
Vulcanodon	Volcano Tooth Lizard	about 180	Africa	20/6.5

*Dinosaur origin/nature has been debated since the first fossil was found in 1861.

SAURISCHIAN DINOSAURS

How are the **saurischian** dinosaurs **classified**?

Paleontologists divide the saurischian dinosaurs into two general groups: the sauropods and the carnivorous theropods (Theropoda). Both of these general groups contained numerous diverse species, which led to further subdivisions of these groups.

Like all classifications, there are differences from system to system, and there are continual debates and changes within each classification scheme. For example, in some classifications, the saurischians are divided into the theropods and the sauropodmorphs; the sauropodmorphs are further divided into the sauropods and prosauropods. But since the prosauropods died out early in the Jurassic period, other classifications concentrate mainly on the sauropods.

The general classification system used here is only one version: The sauropods in the Jurassic were subdivided into the diplodocids (Diplodocidae) and the brachiosaurids (Brachiosauridae); the camarasaurids (Camarasauridae) are either added as a subdivision of the brachiosaurs or are sometimes considered to be a separate group.

The theropods were subdivided into the ceratosaurs (Ceratosauria) and the tetanurans (Tetanurae); the tetanurans were further divided into the coelurosaurs (Coelurosauria) and carnosaurs (Carnosauria)—also known as the allosaurs (Allosauria). Some classification systems further divide the coelurosaurs into the ornithomimosaurs (Ornithomimids) and the maniraptorans (Maniraptora).

What were some **general characteristics** of the **sauropods**?

The sauropods were quadruped animal that ranged from the relatively small (approximately 23 feet [7 meters] long) to the longest land animals ever known (up to 131 feet [40 meters] long; although it is thought that some sauropods reached 180 to 200 feet [55 to 60 meters] in length). They had very long tails and necks, five-toed "hands" and feet, massive limbs, very small heads, and peg-like teeth. Sauropods were herbivorous, and had the lowest ratio of brain-to-body mass (called the encephalization quotient) of all the dinosaurs. These animals first appeared in the Early Jurassic period; by the Late Jurassic period, they had reached the pinnacle of their evolution and diversity.

In general, what **characteristics** determine if a dinosaur **skeleton** is a **sauropod**?

Paleontologists use several technical characteristics to determine if a dinosaur skeleton is that of a sauropod. In general, they look for the following:

1. Twelve or more neck (cervical) vertebrae.
2. Four or more sacral (between the hipbones) vertebrae (most modern reptiles have two; birds have over ten; and most modern mammals have three to five).
3. Massive, vertical limbs with long, solid bones.
4. Ilium (part of the pelvis bone) expanded to the back.

All sauropods have extra neck vertebrae, which is an evolutionary feature that developed at the expense of the back (dorsal) vertebrae. In addition, their skulls were weakly attached; thus the skull is often missing from the rest of the fossil skeleton. Additional skeletal characteristics are often used to classify fossils into groups and species, including the tail chevrons and the socket structure between the vertebrae.

What were the **prosauropods**?

The prosauropods (Prosauropoda) were some of the first dinosaurs to be discovered and described in the 1830s, which was even before the term dinosaur was coined to describe these huge reptiles. The name prosauropod, or "precursor of the sauropods," is a misnomer, as the earliest known creatures were already too specialized to be the ancestors of the sauropods, but the name is still used today.

What is the most primitive sauropod known?

The most primitive sauropod known is the *Vulcanodon,* an early Jurassic period dinosaur discovered in Zimbabwe, Africa. Unfortunately, the skull and vertebrae (except for the tail and part of the pelvic bone) have not been found. It is thought that the dinosaur walked on all fours and was about 33 feet (10 meters) long; the bones showed both prosauropod- and sauropod-like features.

The prosauropods evolved during the late Triassic period some 230 million years ago, and they apparently disappeared at the end of the Early Jurassic period. Most prosauropods had blunt teeth, long forelimbs, and extremely large claws on the first finger of the forefoot (often called the thumb claw); most were semi-bipedal (walked on two legs). They were mostly herbivores, and only toward the Early Jurassic gained the huge size or special adaptations of the later herbivorous dinosaurs.

It is currently difficult to pin down the classification of the prosauropods. In certain classifications, saurischian dinosaurs were also divided into the suborder Sauropodomorpha, with another division Prosauropoda. Another classification suggests the Sauropodomorpha were divided into the Sauropoda, Prosauropoda, and Segnosauria. Still other classifications just list the prosauropods as an extinct offshoot of the saurischians.

What **prosauropod fossils** have been **discovered**?

In 1836, not long after the discovery of the first dinosaur fossils (although the term had not yet been coined), one of the earliest fossil prosauropods was found: a *Thecodontosaurus;* most of the fossil was later destroyed during bombings in World War II (since then, others have been discovered). A fossil *Plateosaurus* was found in 1834, thought to be the largest of its time and today considered the best-known and most extensively studied of all the prosauropods. Other prosauropod fossils have been discovered over the years, including the *Azendohsaurus, Sellosaurus, Saturnalia,* and *Riojasaurus*—all from the Late Triassic period. From the Early Jurassic period, the most common prosauropods were the *Massospondylus, Yunannosaurus,* and *Lufengosaurus*.

What were some **major groups of sauropods** during the **Jurassic**?

There were several major groups of sauropods during the Jurassic period. They included the following:

Cetiosaurids (Cetiosauridae): The cetiosaurs were a group of early sauropods. In fact, this group was an amalgamation of many different types of sauropods that had relatively simple vertebrae. Although some lived until the Late Jurassic period, the dinosaurs in this loose group retained some primitive features. Scientists believe that the cetiosaurids led the way, evolutionarily, for the more advanced forms of sauropods, such as the diplodocids, brachiosaurs, and camarasaurs. Some well preserved specimens of this group have recently been found in China, complete with skulls. These sauropods were distinguished by the lack of an opening in the jaw and five pelvic vertebrae, among other features. The cetiosaurids include the Middle Jurassic *Shunosaurus,* with a tail that ended in a club of bone; and the Late Jurassic *Mamenchisaurus,* with an extremely long neck containing 19 elongated vertebrae. The *Mamenchisaurus* was close to diplodocid line, but still retained a number of primitive characteristics that distinguished it from them.

Diplodocids (Diplodocidae): The diplodocids were a group of advanced sauropods living in the Late Jurassic period; they included some of the longest-known dinosaurs. The diplodocids had long, whip-like tails with at least 80 vertebrae, vertebrae with tall spines, small, long, and slender skulls with elongated muzzles, peg-like teeth only located in the front of the mouth, and nostrils on top of the head.

Brachiosaurids (Brachiosauridae): The brachiosaurids were another group of sauropods. These Jurassic period dinosaurs were much more massive than the diplodocids. Their most unique characteristic was front legs as long as—or even longer than—their rear legs. This, combined with their long necks, gave them a giraffe-like posture. In addition, the brachiosaurids had a relatively short tail comprised of about 50 small vertebrae, nostrils perched on a protrusion on top of the head, and a long neck with an average of 13 large vertebrae.

Camarasaurids (Camarasauridae): Depending on the classification system, the sauropods called camarasaurids can be listed as part of the brachiosaurids or placed in a group of their own. They were shorter and heavier than the diplodocids, with front and rear legs more similar in length. These Late Jurassic (to Late Cretaceous) period dinosaurs were the largest vegetarians had an average of 12 neck vertebrae; low, thick spines; vertebrae with extensive, deep cavities; large nostrils in front of the eyes; and large, spoon-like teeth set in a short, blunt skull. The *Camarasaurus* is the most common sauropod—and one of the smallest measuring 30 to 60 feet (9 to 18 meters) in the Morrison formation in the western United States. In fact, it is one of the only dinosaurs whose osteology—or the anatomy and structure of the bones—is completely known.

What were some **Jurassic period sauropods**?

The following lists the most prominent Jurassic period sauropods and some of their characteristics:

Diplodocidae

Diplodocus: This dinosaur gives this group its name; it was 89 feet (27 meters) in length, with an estimated weight of 11 to 12 tons (10 to 11 metric tons).

Apatosaurus: Once commonly known as *Brontosaurus,* it was shorter and stockier than the *Diplodocus*; it was also one of the largest land animals that ever existed, measuring over 75 feet (23 meters).

Barosaurus: Similar to *Diplodocus,* but its cervical vertebrae were 33 percent longer.

The *Diplodocus* was nearly 90 feet (27 meters) long, weighing in at over 10 tons (9 metric tons), and was one of the largest sauropods to roam Earth (iStock).

Suuwassea: Found in the Morrison formation of the United States, it measured about 46 to 49 feet (14 to 15 meters) in length.

Supersaurus: Another candidate for the longest dinosaur; it is estimated to have been about 108 to 112 feet (33 to 34 meters) long.

Brachiosauridae

Brachiosaurus: This dinosaur gives its name to the group; the largest found fossil of this late Jurassic dinosaur was approximately 82 feet (25 meters) long and 42 feet (13 meters) high, about the height of a four-story building.

Lusotitan: This dinosaur had long forearms; it also averaged about 82 feet (25 meters) long. The most complete fossil found in Portugal did not have a skull, but scientists still believe it was a member of the Brachiosauridae.

Camarasauridae

Camarasaurus: A relatively small sauropod; it was approximately 60 feet (18 meters) long; its forelimbs were not as proportionally long as the brachiosaurids.

What were some **general characteristics** of the other saurischian group, the **theropods**?

The theropods (Theropoda) had many general characteristics. Because they were carnivores, their teeth tended to be blade-like, with serrated ridges. Their claws, especially on the hands, were often recurved and tapered to sharp points. The outer fingers of the hand were either reduced in size or were completely lost. Most theropods were long-legged, bipedal, slender, and quick—all characteristics that enabled the animals to more easily catch their prey. They also had three walking toes on the hind feet.

Theropods had hollow limb bones; some went one step further and had air-filled (pneumatic) bones in certain parts of their body. For example, in some dinosaurs, pneumatic bones were found in the middle of the tail; in others, the air-filled bones were present at the back of the skull.

There is a good deal of evidence that larger theropods attacked other animals. For example, sauropod bones found in Colorado show the bite marks from a large theropod—and sometimes the bones indicate that the prey survived the attack. At other sites, trackways show the footprints of theropods chasing or stalking smaller sauropods.

In general, what characteristics determine if a **dinosaur skeleton** is that of a **theropod**?

There are many unique skeletal characteristics paleontologists use to determine if certain dinosaur bones belong to a theropod. In general, here are a few of the characteristics:

1. The bone in front of the eye (lacrimal) extends onto the top of the skull.
2. The lower jaw has an extra joint.
3. The shoulder blade (scapula) is strap-like.
4. The upper arm bone (humerus) is less than half the length of the upper leg bone (femur).
5. The elongated hand has lost (or has shrunken) two outer fingers.
6. The top of the bones in the palm of the hand (metacarpals) has pits where ligaments were attached.
7. The fingers have elongated bones between the second-to-last and last joints.
8. Near the head of the upper leg bone (femur) is a shelf-like ridge for the attachment of muscles.

How are the **theropods** classified?

Like most dinosaur science, the classification of theropods differs greatly from study to study—and even scientist to scientist. One common classification lists the following: Suborder Theropoda, Infraorder Ceratosauria, Clade Tetanurae, Infraorder Carnosauria, Clade Coelurosauria, Infraorder Ornithomimosauria, Clade Maniraptora, and so on. As more dinosaurs are found, this listing continues to change.

What were the **ceratosaurs**?

The ceratosaurs (Ceratosauria, or "horned reptiles") comprise a major theropod division. These were among the earliest theropods, arising during the Triassic period and evolving into much larger animals during the Jurassic period.

Allosaurs were a type of tetanuran, which included other carnivores such as velociraptors and the tyrannosaurs (iStock).

Similar to all known theropods, the ceratosaurs had hollow bones—bones that were much stronger and easier to bend than solid ones. The ceratosaur dinosaurs had strongly curved, S-shaped necks, similar to those exhibited by modern birds, but on a much larger scale. (Scientifically, this is expressed by saying the ceratosaurs began to exhibit bird-like features, or more appropriately, birds show ceratosaurian features.)

Another ceratosaur characteristic included its upper jaw bone structure: there was a loose attachment between the two upper jaw bones, known as the premaxilla (a front bone of the upper jaw) and the maxilla (main upper jaw bone, behind the premaxilla). The notch formed by this arrangement was filled by a large tooth in the lower jaw when the mouth was closed.

What were the **tetanurans**?

The tetanurans (Tetanurae, or "stiff tail") are regarded as a sister group of the ceratosaurs; they are considered to be all theropods more closely related to modern birds than to the ceratosaurs. The tetanurans were a large and diverse group, comprising many species, including most of the well-known theropods. The unique characteristics of this group include a special row of teeth, a large pubic boot, the presence of a large opening in front of the jaw, and a rear half of the tail stiffened by interlocking, rod-like projections of vertebrae. The tetanurans also had a three-fingered hand. Many popular dinosaurs are tetanurans, including *Archaeopteryx, Allosaurus, Oviraptor, Spinosaurus, Tyrannosaurus, Velociraptor,* and all species of modern bird.

What were the **carnosaurs**?

The carnosaurs (Carnosauria, Greek for "meat-eating lizards") first appeared in the Middle Jurassic (some say the Late Jurassic) period and lasted until the end of the Cretaceous period. They included the families of Sinraptoridae, Allosauridae, and Carcharodontosauridae, and currently encompass only the allosaurs and their closest kin.

The dinosaurs that made up the carnosaurs were large, heavy predators. In addition to being large in size, these dinosaurs shared other unique characteristics, such as a large cavity in the lacrimal bone of the skull; this cavity, located in front of and above the orbit (eye opening in skull), may have held a gland. Other characteristics included large eye openings in a long, narrow skull, femurs (thigh bones) larger than the tibia (shin bone), and neck vertebrae with a ball joint on the front and a socket on the back. Carnosaurs also had fairly large forelimbs. Some very large carnosaurs in the carcharodontosaurid family (a clade [class] within the Carnisauria), such as *Giganotosaurus* and *Tyrannotitan,* are among the largest known predatory dinosaurs ever known.

What were the **coelurosaurs**?

The coelurosaurs (Coelurosauria) make up the clade containing all theropod dinosaurs more closely related to birds than to carnosaurs. It is a diverse group that includes tyrannosaurs, ornithomimosaurs, and maniraptors; Maniraptora includes birds, the only descendants of coelurosaurs alive today. In the past, this classification included all the small-bodied theropods, while the carnosaurs included all the large-bodied theropods. This is not the case today—and although several alternate classifications have been proposed, none is universally accepted.

Because this group of theropods was extremely diverse, the group's definition, the dinosaurs within the group, and the relationships between the various species are currently in a continual state of change. Presently, the major subgroups of coelurosaurs are the ornithomimosaurs (Ornithomimids), maniraptorans (Maniraptora), and a fairly recent addition to the coelurosaurs, the tyrannosaurs. Some of the dinosaurs in this group evolved during the Late Jurassic period, but the majority reached their peak in the Cretaceous period.

These carnivorous dinosaurs were more bird-like in characteristics and appearance than the large carnosaurs. In fact, birds are classified in the coelurosaur subgroup known as the maniraptorans (Maniraptora). Coelurosaurs had very long forelimbs and well-developed, hinge-like ankles, although later dinosaurs in this group may have lost these features or had modified versions. Some other characteristics of the coelurosaurs include some special bone structures, such as a triangular bulge on the lower part of the pelvis and a protrusion on the ankle bone.

Even if they aren't direct ancestors of our modern day ostriches, the ornithomimosaurs looked very much like them (iStock).

The tyrannosaurs (also referred to as tyrannosaurids, depending on the text), including *Tyrannosaurus rex,* used to be classified by most scientists as carnosaurs. But recent cladistic (class) analysis shows that the tyrannosaurs are more closely related to the coelurosaurs. (Coelurosaurs are a class of dinosaurs related to birds; tyrannosaurs also belong to the larger group known as tetanurans, dinosaurs with "stiff tails.") Although some probably evolved during the Late Jurassic period, most of them dominated the Cretaceous period.

What were the **ornithomimosaurs**?

The ornithomimosaurs (members of the clade Ornithomimosauria, or "bird mimics") were a theropod dinosaur that—from general superficial appearances—resembled modern-day ratite birds such as ostriches. These dinosaurs were slender, with flexible necks, small heads, toothless beaks, long slim hind limbs, elongated arms, and grasping hands with three powerful fingers. Ornithomimosaurs arose in the Late Jurassic period and died out at the end of the Cretaceous.

These dinosaurs were not limited to small sizes; some could reach up to 20 feet (6 meters) in length. Because of their lower leg bones and foot structure, some scientists think that these dinosaurs, like the ostrich, could run very fast. Estimates for one species, the *Struthiomimus,* range from 22 to 37 miles (35 to 60 kilometer) per hour. Other ornithomimosaurs include *Gallimimus, Anserimimus,* and *Ornithomimus*.

What were the **maniraptorans**?

The maniraptorans (Maniraptora, or "hand snatchers") were a clade of coelurosaurian dinosaurs that included the birds and dinosaurs most closely related to birds. They are defined as all dinosaurs closer to birds than the ornithomimosaurs. In fact, many paleontologists believe birds came from maniraptorans, evolving from this group during the Jurassic period.

The maniraptorans were a very diverse group and looked at first glance to be totally unrelated. However, they all shared some common characteristics that made them part of this group. The maniraptorans included the dromaeosaurids, troodontids, therizinosaurs (or segnosaurs), oviraptorosaurs, and, more recently, the aves (birds).

All of the maniraptorans shared the following general characteristics: a curved bone (semilunate carpal) in the wrist, a fused collar bone (clavicle) and breast bone (sternum), long arms, a hand (manus) larger than the foot (pes), a pubis (pelvic bone) that pointed downwards, and a shortened tail that was stiffened toward the end.

Were there any **dromaeosaurs, troodontids, therizinosaurs**, or **oviraptorosaurs** in the **Jurassic period**?

There is little evidence of dromaeosaur, troodontid, therizinosaur (or segnosaur), and oviraptorosaur fossils in Jurassic period rock, although dromaeosaurid and troodontid teeth have been reported in late Jurassic rock. It is currently thought that these creatures evolved in the Late Jurassic, and were very prolific in the Cretaceous period.

What were some **Jurassic period theropods**?

The following lists the some of the Jurassic period theropods and their characteristics:

Ceratosauria

Ceratosaurus: A large, heavy carnivorous dinosaur up to 23 feet (7 meters) long; it had short blade-like crests over the eyes, and a short triangular nose horn; it was the largest known ceratosaur, and appeared in the late Jurassic period; some scientists believe the dinosaur was an oddball: it appeared long after the other ceratosaurs, and may not even belong to this group; in fact, some people put it with the carnosaurs.

Dilophosauridae

Dilophosaurus: Up to 20 feet (6 meters) long and fairly slender; the skull had a double crest of thin, parallel plates on its nose and forehead; it was fancifully portrayed in the movie *Jurassic Park* with a frill, spitting poison, and much smaller than in actuality.

Carnosauria

Allosaurus: A large predator, with fairly long and well-muscled forelimbs and huge claws; it had large legs with heavy, clawed feet, and large, narrow jaws.

Megalosaurus: The first dinosaur to be described; it was thought to be up to 30 feet (9 meters) long, but recently studies of the fragmentary remains indicate that they may actually be parts of different carnosaurs.

The *Dilophosaurus* is the only member species of the Dilophosauridae. It was distinctive for the double crest on its head, but the idea that it could spit poison, as portrayed in the film *Jurassic Park,* is a Hollywood invention (Big Stock Photo).

Coelurosauria

Megapnosaurus: For "big dead lizard," this small and slender dinosaur measuring 10 feet (3 meters) long, and similar in form to *Coelophysis,* was once known as *Syntarsus*.

Compsognathus: A small coelurosaur that evolved during the Late Jurassic; *Compsognathus longipes* is the only recognized species; it was found in a limestone quarry in Solnhofen, Germany, where the first known bird, *Archaeopteryx lithographica,* was also discovered.

Tyrannosaurus rex: The *T. rex,* or "tyrant lizard," was an extremely large carnivorous dinosaur that evolved during the Late Jurassic period; it became the dominant predator of its time, growing close to 50 feet (15 meters) long.

Ornithomimosauria

Gallimimus: An ornithomimosaur that evolved during the Late Jurassic; the *Gallimimus bullatus* was featured in the movie *Jurassic Park,* and was thought to have been one of the fastest dinosaur runners, judging from its long hind limbs.

ORNITHISCHIAN DINOSAURS

How are the **ornithischian dinosaurs classified**?

The ornithischian dinosaurs, except for some early primitive forms, are all members of a clade called the Genasauria: beaked, herbivorous dinosaurs that were the prey for carnivorous theropods. In one frequently used classification, the Genasauria are divided into the Neornithischia and Thyreophora. The Thyreophora are further divided into the Stegosauria and the Ankylosauria (although in some classifications, these two divisions are combined into one group; in another classification, Eurypoda is added with the Stegosauria and Ankylosauria); the Neornithischia are divided into several types of dinosaurs, the major one being the Ceropoda, which include the Ornithopoda and the Marginocephalia. All the Genasauria dinosaurs developed in the Middle to Late Triassic period, evolving and increasing in number by the Jurassic and Cretaceous periods.

All ornithischians have a few major features in common: a pubis pointing backward, running parallel with the ischium (the name *Ornithischia* means "bird-hipped," and modern birds also have pelvises in which the pubis points backwards). Besides the hip structure of the animals, the other main characteristic shared by the ornithischian (Genasauria) dinosaurs was a deeply inset tooth row. This manifested itself in the skull as a deep concavity on each side, a structure some paleontologists use to support the idea of cheeks in some of these dinosaurs. (Since all ornithischian dinosaurs were herbivorous, cheeks would have been an important feature for keeping the food in the mouth during the chewing process.)

What were two of the **earliest known ornithischian** dinosaurs?

Two of the earliest known ornithischian dinosaurs were *Lesothosaurus* and *Heterodontosaurus,* both of whose remains were discovered in Late Triassic to Early Jurassic rock of the Stormberg Group in southern Africa. The *Lesothosaurus* was a small, slender biped about three feet (one meter) long, with long hind legs, delicate, short arms and a five-fingered hand. Although this dinosaur had the characteristic ornithischian hip structure, it lacked a deeply inset tooth row, meaning its cheeks were not well developed. Therefore, paleontologists suggest that the *Lesothosaurus* was the most primitive known ornithischian dinosaur; it is often associated with the Genasauria division, Neornithischia.

The *Heterodontosaurus,* another dinosaur about three feet (one meter) long, differed from *Lesothosaurus* in the skull and hands. Its skull had a deeply inset tooth row, the tooth pattern was more complicated, and it had thicker outside enamel on the upper and inside of the lower teeth. The *Heterodontosaurus* was more advanced than *Lesothosaurus,* exhibiting characteristics some shared by later ornithischian dinosaurs.

What are the main **characteristics** of the **thyreophorans**?

The thyreophorans (Thyreophora, or "shield bearers") were one of the two main Genasauria dinosaur divisions. This group was characterized by the presence of bony

With bony armor being the main characteristic of thyreophorans, the *Stegosaurus* makes a perfect example of the group (iStock).

armor on their bodies. The thyreophorans were further subdivided based on their type of armor: The stegosaurs (Stegosauria, or "plated dinosaurs") had armor manifested as bony plates or spines on their backs; the ankylosaurs (Ankylosauria, or "crooked or bent reptiles") had armor as a covering, or small plates over their backs. In some classifications, there is no division between the stegosaurs and ankylosaurs.

What were some **early thyreophorans**?

Some early thyreophorans were the *Scutellosaurus* and *Scelidosaurus*. The *Scutellosaurus*—an Early Jurassic period dinosaur whose remains were discovered in western North America—was small and bipedal (two-footed), similar to the *Lesothosaurus*. The *Scutellosaurus* had numerous small bony skin plates (although not as large as later thyreophorans) and was among the smallest of the armored dinosaurs, growing from just over a foot (around a half meter) to 3 feet (1 meter) in length. The *Scelidosaurus* was an Early Jurassic period dinosaur living in western Europe and England about 180 million years ago. It was a quadruped (four-footed) and was much heavier than *Scutellosaurus,* growing to a length of approximately 13 feet (4 meters). It had heavy, bony plates, and hoof-like claws.

What were the **stegosaurs**?

The stegosaurs (Stegosauria) were one of the two groups comprising the thyreophorans (although some classifications group them with the anklyosaurs in the thyreophorans division). These quadruped, "plated" dinosaurs first appeared in the Middle Jurassic. They were medium in size, from 13 to 26 feet (4 to 9 meters) long, with a small, long skull with simple teeth, and a long toothless beak. They also had armor consisting of two rows of vertical bony plates or spines along their tail, back, and neck. Fossils of stegosaurs—each species with a unique arrangement of spines and plates—are found in Jurassic and Cretaceous periods rock layers around the world, but they were most abundant in the Jurassic.

Why were **stegosaurs** originally thought to have had a **"second brain"**?

The dinosaurs that made up the Stegosauria, such as *Stegosaurus* and *Kentrosaurus,* had extremely tiny brains. For example, the *Stegosaurus,* an animal that weighed approximately 1 to 2 tons (0.9 to 1.8 metric tons), had a brain only weighing 0.15 to 0.18 pounds (70 to 80 grams) and thought to be about the size of a walnut. Paleontologists also found an enlarged part of spinal cord in fossils of the stegosaur sacrum. This swelling, larger than the brain cavity, was originally thought to be a "second brain." Currently, scientists believe this area is associated with nervous and connective tissue and fat, not a second brain.

What were the **ankylosaurs**?

The ankylosaurs (Ankylosauria) were one of the two divisions that made up the thyreophorans (the stegosaur was the other). Paleontologists divide the ankylosauria into two more families—the Nodosaurids and Ankylosaurids—based on differences in skulls, shoulder blades, and armor.

The ankylosaurs were medium-sized quadruped dinosaurs found in many parts of the world. In fact, the first dinosaur ever discovered in Antarctica was the ankylosaurian *Antarctopelta;* the fossils were recovered from Ross Island in 1986. Ankylosaurs were short-legged and squat, with long, wide bodies. Their heads ranged from long and narrow to wide with broad muzzles. All of these dinosaurs had bony plates of armor over their bodies, often with spines, spikes, or studs projecting outward. Some even had bony clubs on the ends of their tails. Armor consisted of plates of bone, known as scutes, embedded in the skin; some dinosaurs even had their heads covered with this armor. They were the "tanks" of the dinosaur world and were able to fight off almost any predator.

Unfortunately, very little is known about these dinosaurs, since so few complete fossils have been found. The ankylosaurs first appeared during the Early to Middle Jurassic periods, but they became most numerous and diverse during the Cretaceous period. Most of the ankylosaurs belonged to the Nodosaurid subgroup early in the Cretaceous period (although one genus, *Sarcolestes,* is from the Jurassic); the subgroup Ankylosaurid were more prevalent in the latter Cretaceous period, and were distinguished by their broad heads, spikes in the back of their skulls, and club-like tails.

What were the **cerapods**?

The cerapods (Cerapoda, or "horn-footed") were the group within the Neornithischia—a sister group of the Thyreophora within the clade Genasauria. They are divided into the ornithopods (Ornithopoda; "bird-foot") and marginocephalians (Marginocephalia; "fringed-heads"); other classifications divide the Margin-

Hadrosaurs, with their prominent crests, are one of the most easily recognizable dinosaurs of the Jurassic period (iStock).

ocephalia into two groups: the Pachycephalosauria ("thick-headed lizards") and Ceratopsia ("horned-face"). In general, the primary characteristic of the cerapods was that they had only five or fewer premaxillary teeth, with the enamel distributed differently on the sides of their teeth. The Marginocephalia are often distinguished by a bony shelf they had at the back of the skull. One example is the *Stormbergia* dinosaur fossil from the Early Jurassic period in South Africa.

What **ornithopod dinosaurs** lived in the **Jurassic** period?

One of the most well-known—and earliest discovered dinosaurs—was the *Iguanodon* ("iguana tooth"). Ornithopods also included the hadrosaurs (the "duck-billed" dinosaurs with the famous crests), the iguanodontids, the heterodontosaurs, the hypsilophodontids, and numerous other dinosaurs. But many of these dinosaurs were not prolific until the Cretaceous period. In general, the ornithopods were bipedal, ranging in size from less than 3 feet (1 meter) tall and six feet (two meters) long, to as big as about 23 feet (7 meters) tall and 66 feet (20 meters) long. The ornithopods were plant eaters, evolving early in the Jurassic period and continuing to the end of the Cretaceous. They lived on every continent, including Antarctica. They were the first herbivorous dinosaurs to have multiple tooth rows, cheek pouches, and the ability to truly chew.

What were some **Jurassic period ornithischian dinosaurs**?

The following are some examples of the ornithischian dinosaurs that made up this group during the Jurassic period:

Thyreophora

Stegosauria

Huayangosaurus: Found in China during the Middle Jurassic, these animals had small plates in skin, with spike-like armor and equal length front and rear legs; they were approximately 13 feet (4 meters) long, with short snouts; they are considered the most primitive of the stegosaurs.

Stegosaurus: A Late Jurassic dinosaur weighing approximately one to two tons; it had an array of bony plates along the length of the back and tail spikes; the hind legs were long, and the forelegs were short and massive; the head was small and elongated, and the brain size was extremely small for an animal of this size. More than 80 specimens have been uncovered in the United States' Morrison formation.

Kentrosaurus: This dinosaur (the name means "pointed lizard") had spines on its tail, hip, shoulder, and back; its bony plates, similar to those on *Stegosaurus,* were also present on the neck and anterior part of the back. It measured close to 13 feet (4 meters) long and may have been more flexible than its close relative, the *Stegosaurus*.

Ankylosauria

Sarcolestes: This dinosaur ("flesh robber") developed in the Middle Jurassic period, and is thought to be one of the earliest ankylosaurs. It had a large piece of armor-plating on its outer surface and when found in the late 1800s, was first thought to be a flesh-eating dinosaur—thus its name.

Dracopelta: The *Dracopelta* ("armored dragon") was a small, Late Jurassic period ankylosaur from Portugal. There are few fossil remains, thus its size is only estimated to be around 6 feet (2 meters) in length.

Neornithischia

Ornithopoda

Camptosaurus: Called "bent lizard," this beaked dinosaur was a medium-sized, bipedal herbivore that lived in the Late Jurassic to Early Cretaceous periods. It reached lengths up to 26 feet (7.9 meters) and was 6.7 feet (2 meters) tall at the hip. It is also thought to be the ancestor to many of the highly successful, plant-eating dinosaurs in the Cretaceous period.

Heterodontosaurus: This dinosaur was only about 3 feet (1 meter) in length. It had canine-like teeth and relatively long arms, with large hands; its teeth were designed for cutting.

Marginocephalia

Ceratops: This dinosaur (meaning "horn face") was a ceratopsian that lived in the Late Cretaceous period. Fossils of this creature are found in the United States in Montana, as well as in Alberta, Canada. It is famous for the horn that sticks out from the middle of its "forehead" and above the nose.

GENERAL JURASSIC DINOSAUR FACTS

Which major **Jurassic period dinosaurs** were **herbivores** and **carnivores**?

The largest dinosaurs in the Jurassic tended to be plant eaters. The best known examples include the long-necked sauropods like *Brachiosaurus* and *Apatosaurus,* creatures that ate leaves off the tops of high trees. In addition, all of the ornithischians were herbivores, such as the plated *Stegosaurus* and the ankylosaurs. Ornithopods such as *Camptosaurus* competed with the large sauropods for vegetation. In contrast, the theropods were all carnivores, and evolved into large predators, such as the *Megalosaurus* and the well-known *Allosaurus*.

What were some of the **smallest and largest dinosaurs** known in the **Jurassic period**?

Some of the dinosaurs living during the Jurassic period evolved into the largest creatures ever to live on land—with the majority being plant eaters. And it seems as if every year brings a new fossil discovery that unearths another, larger dinosaur.

The largest complete dinosaur so far discovered is the *Brachiosaurus,* which measured 75 feet (23 meters) in length and 46 feet (14 meters) in height, or about the height of a four-story building. But this Jurassic dinosaur has yet to win the ultimate "big" prize. The reason for this hesitation in giving the *Brachiosaurus* the "largest" status is familiar in dinosaur studies: there are too few skeletal remains of larger dinosaur species, with only fragmentary leg bones and vertebrae. This makes determining the size of these dinosaur species difficult. For example, many fragments point to such possible larger dinosaurs as the *Argentinasaurus* and *Amphicoelias,* both of which may have been one and a half to two times larger than *Brachiosaurus*.

The two longest dinosaurs in the Jurassic period may have been the plant eaters, the sauropods *Diplodocus* and *Supersaurus*. The *Diplodocus* measured more than 90 feet (27 meters) in length; the *Supersaurus,* thought to be a major contender as the longest land animal ever on Earth, measured close to 138 feet (42 meters) long and 54 feet (16.5 meters) tall. At one time, scientists believed that these huge creatures had to live in the water to support their great bulk; but the latest research suggests that the dinosaurs were able to carry their weight on land.

The smallest Jurassic dinosaurs, called the *Compsognathus* ("pretty jaw"), were just slightly larger than a chicken. This dinosaur was 3 feet (1 meter) long and probably weighed about 6.5 pounds (2.9 kilograms).

Larger carnivorous dinosaurs included the *Allosaurus,* which measured about 50 feet (15 meters) in length. Scientists believe the attack of the *Allosaurus* was amazing: it would open its mouth to the furthest extent, running headlong into its victim. Its 60 curved, dagger-like teeth would plunge into its prey, driven by two tons of dinosaur.

Did the first **dinosaur fossil to be named** come from the **Jurassic period**?

Yes, the first dinosaur fossil to be named was the *Megalosaurus,* named by geologist William Buckland (1784–1856) in 1824. The fossil came from the Middle Jurassic, and was found in Oxfordshire, England.

Once commonly referred to as the *Brontosaurus,* this long-necked sauropod is now more correctly referred to as the *Apatosaurus* (iStock).

What possible large **Jurassic dinosaur** is **only known from drawings** of a **fossil**?

The *Amphicoelias* ("doubly hollow") was a herbivorous sauropod dinosaur that included what may be the largest dinosaur ever discovered—A. *fragillimus* from the Late Jurassic period. The evidence—a single fossil bone—was uncovered in the 1870s. If the description is true, it may have been the longest known dinosaur vertebrae, measuring 131 to 196 feet (40 to 60 meters) in length, with the animal weighing in at 135 tons (122 metric tons); thus rivaling the blue whale, the heaviest animal ever known. However, because the only fossil remains were lost at some point after being studied, the evidence survives only in drawings and field notes.

What **Jurassic period** dinosaur was **formerly** called ***Brontosaurus***?

The *Apatosaurus* was formerly known as the *Brontosaurus*. Fossils from the *Apatosaurus* were officially named in 1877, while the *Brontosaurus* fossils were named in 1878. It wasn't until later that it was noticed that the fossils of the two dinosaurs were really the same. Since the *Apatosaurus* had been named first, it was adopted as the official designation for this animal.

What **Jurassic period** dinosaurs **lived the longest**?

The herbivorous ornithopods lived the longest, from the Early Jurassic to the Late Cretaceous period. They include a series of successively larger and more massive dinosaurs that spread throughout most of the continents. One was the *Heterodon-*

tosaurus of the Early Jurassic period, a quick, 4-foot (1.3-meter) dinosaur, with strong front canine tusks and flexible hands used for digging and gasping vegetation.

OTHER LIFE IN THE JURASSIC

Besides dinosaurs, what **other land animals** were present during the **Jurassic period**?

Although the dinosaurs were the dominant animals, there were other animals that existed during the Jurassic period. Because of the relative stability of the climate and the lush vegetation, many land animals diversified and increased in numbers. But not all creatures survived through the Jurassic period; many became extinct, probably because competition for food increased. Below are short descriptions of the types of animals that existed at the time:

Amphibians

Frogs and salamanders: First modern frogs and salamanders appeared.

Reptiles

Turtles: First modern turtles appeared in the Early Jurassic period; they were able to retract their heads into their upper shell.

Lizards: First true lizards appeared in the Middle Jurassic period.

Crocodylians: First true crocodylians appear, and were small, three-foot- (one-meter-) long reptiles that walked on all fours; they had longer hind legs, indicating that their ancestors were bipedal.

Therapsids: Few families of therapsids, or mammal-like reptiles, lived into the Middle Jurassic period; the anomodonts and therocephalians no longer existed, but the cynodonts did survive into the Middle Jurassic.

Mammals

Small mammals: Small mammals became more profuse and diverse during the Jurassic period; they were still very small animals, about the size of a mouse or rat, with the largest the size of a cat; they were mostly nocturnal.

Triconodonts: Late Triassic to Late Cretaceous mammals; one of the oldest fossil mammals; three cusps of teeth in a straight row give them their name.

Haramyoids: Late Triassic to Middle Jurassic mammals; one of the oldest fossil mammals; their teeth had many cusps in at least two parallel rows.

Symmetrodonts: Late Jurassic to Early Cretaceous mammals; they had upper and lower cheek teeth with many cusps in a triangular pattern.

Docodonts: Middle to Late Jurassic period mammals; they had elaborate cheek teeth, with most of the cusps in a T-shape.

A reef shark swims near the Bahamas. As far back as the Jurassic period, sharks swam in Earth's oceans and were some of the deadlist predators of the deep (iStock).

Multituberculates: The multituberculates were the largest group of mammals in the Mesozoic, first appearing in the Late Jurassic period.

What were the **major marine animals** living during the **Jurassic**?

There were many marine animals that thrived during the Jurassic period, ranging from small seabed dwellers to large swimming predators. Most of these animals were similar to those found in the Triassic period, although many had diversified and increased in number during the Jurassic period.

Modern shark families developed at this time; bony fishes with symmetrical tails, the teleosts, diversified (they account for the great majority of modern fishes—over 20,000 species). The first oysters evolved; modern squids and cuttlefishes appeared; and squid-like belemnites diversified. The ammonoids almost disappeared during the late Triassic extinctions; one family survived (out of eight) and quickly diversified during the Jurassic period.

The ichthyosaurs, such as the *Ichthyosaurus* and *Stenopterygius,* flourished in the Jurassic period oceans. Plesiosaurs were also abundant, such as the *Muraenosaurus,* a Late Jurassic period reptile with a 36-foot (11-meter) neck (it included 40 vertebrae); and the *Liopleurodon,* a short-necked pliosaur with a huge, elongated head was 39 feet (12 meters) long, with a 10-foot (3-meter) head.

What was so **special** about the evolution of the **ichthyosaurs**?

Ichthyosaurs were streamlined, dolphin-shaped reptiles, but their backgrounds differed from many reptiles living around them: These creatures' ancestors went back to the sea after living on the land. Some scientists believe that ichthyosaurs were the first major group of reptiles to return to the sea. At first, they no doubt stayed close to the shoreline, just as seals and walruses do today. But after millions of years, the creatures went into the oceans, spread, and eventually became totally fish-shaped.

The oldest ichthyosaur fossils, and the most primitive so far, are 240-million-year-old fossils found in Japan. The primitive ichthyosaur, measuring about 9 feet (2.7 meters) long, probably lived its entire life in the water. Its shape was not yet like a dolphin; and its pelvis bone was still attached to the vertebrae, similar to those of land animals and dissimilar to the later ichthyosaurs. The primitive ichthyosaurs also had fins that were similar to the limbs of land reptiles (with splayed fingers). In other words, this primitive fossil shows the first step the ichthyosaurs took from the land to the oceans.

There was one other strange characteristic of the ichthyosaurs: For unknown reasons (as evident in the fossil record), about 135 million years ago, the animals began to fade away, becoming totally extinct between 90 and 100 million years ago. This was much earlier than the demise of the dinosaurs about 65 million years ago.

CRETACEOUS PERIOD

What was the **Cretaceous period** and how did it get its **name**?

The Cretaceous period followed the Jurassic period on the geological time scale; it was the last period in the Mesozoic era. The period lasted from approximately 145 to 65 million years ago, or 80 million years total. The geologic time scale is not exact, and the dates of the Cretaceous period on various scales vary by about 5 to 10 million years. Most of the large Jurassic sauropods, stegosaurs, and theropods disappeared in the early part of this period, but were replaced by an incredibly large array of new dinosaur groups. These included the horned types, duck-billed, and armored sauropods, and new types of theropod carnivores.

In a roundabout way, the Cretaceous period got its name from the type of rock deposited along the northern shores of the Tethys Sea in a band running from what is now Ireland and Britain to the Middle East. This rock—formed from the metamorphosed deposits of the tiny limestone skeletons of diatoms—is known as chalk. The Latin for chalk is *creta,* so the period was named the Cretaceous. "Cretaceous" was first used to describe such a rock found in France. In 1822, Jean-Baptiste-Julien Omalius d'Halloy (1783–1875) used the name "Terrain Cretacé" to describe the strata and associated units of chalk (*craie* in French) found in that country. Since these same strata were also present across the English Channel, English geologists began calling them the Cretaceous System.

What are the major **divisions** of the **Cretaceous**?

Scientists divide the Cretaceous into two epochs: the Early Cretaceous (also called the Lower Cretaceous or, less formally, lower or early Cretaceous), from approximately 144 to 89 million years ago; and the Late Cretaceous (also called the Upper Cretaceous, or upper or late Cretaceous), from approximately 89 to 65 million years ago. Each of these main epochs is broken up into smaller ages. The following chart gives the European nomenclature for each age:

Cretaceous Period

Epoch	Age	Millions of Years Ago (approximate)
Late	Maastrichtian	74–65
	Campanian	83–74
	Santonian	87–83
	Coniacian	89–87
Early	Turonian	93–89
	Cenomanian	97–93
	Albian	112–97
	Aptian	125–112
	Barremian	132–125
	Hauterivian	135–132
	Valanginian	141–135
	Berriasian	144–141

Why did the dinosaurs **thrive and diversify** in the **Cretaceous**?

Although there is no clear answer to this question, paleontologists know that a revolution in life occurred during this time period. This took place as many types of modern flora and fauna made their first appearances. Some scientists theorize that it was the development, and eventual dominance, of a new group of plants, the angiosperms (flowering plants), in harmony with the development of new groups of insects that provided fresh sources of food for the dinosaurs. All these new food sources allowed the dinosaurs to continue to dominate throughout the Cretaceous period.

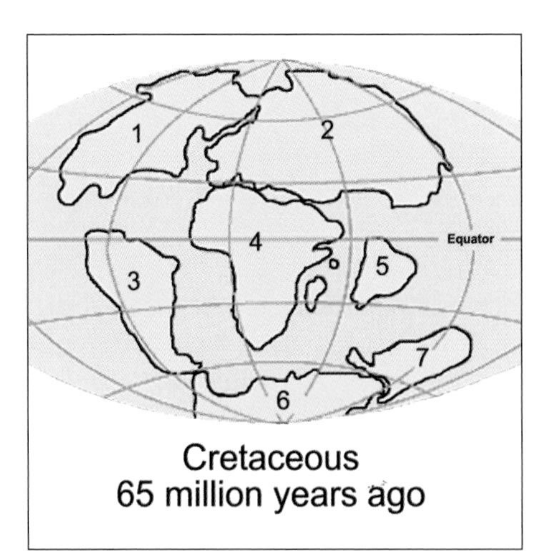

By the Cretaceous period, continental drift had moved the continents to nearly familiar positions: 1) North America, 2) Eurasia, 3) South America, 4) Africa, 5) the Indian subcontinent, 6) Antarctica, and 7) Australia (map based on a U.S. Geological Survey illustration).

How did **Laurasia** and **Gondwana (Gondwanaland)** change during the **Cretaceous period**?

During the Cretaceous period, both Laurasia and Gondwana fragmented into smaller landmasses and separated from each other. The whole motif of the Cretaceous was change—change from the ancient topography to the more familiar forms we see today.

In the Early Cretaceous period, Laurasia began to break up due to the action of an extension of the Mid-Atlantic Ridge, with North America and Greenland separating from Eurasia. Rifting occurred in Gondwana, with South America and Africa beginning to separate. In the middle of the Cretaceous period, Gondwana had separated into four major landmasses: South America, Africa, the combined India and Madagascar, and the combined Antarctica and Australia.

By the Late Cretaceous, North America and Greenland began to split, as did Australia and Antarctica, and India and Madagascar. The Atlantic Ocean continued to widen, and India and Australia moved northward. By the end of the Cretaceous, the continents began to assume their modern outlines and headed toward their current destinations on the planet's surface. This also led to the development and widening of our modern oceans and seas.

IMPORTANT CRETACEOUS DINOSAURS

What were the **major dinosaurs** during the **Cretaceous period**?

During the Early Cretaceous period, many of the Jurassic period dinosaurs disappeared. They were replaced by new, more diverse forms. Toward the end of the Cretaceous, the diversity of dinosaurs dropped dramatically. Of the saurischian sauropods, only the titanosaurids (Late Jurassic or Early Cretaceous period) remained as a major group—and these herbivores were mostly found on the landmasses of Gondwana until the end of the Cretaceous period. In fact, through fossil evidence, it is thought that the titanosaurids replaced other sauropods, like the diplodocids and brachiosaurids, both of which died out around the Late Jurassic to Middle Cretaceous periods. Many of the saurischian theropods also became extinct during the Cretaceous period, while others diversified into a wide range of animals, from large carnivores like *Tyrannosaurus,* to speedy, agile predators like *Velociraptor*.

The ornithischian dinosaurs were the most numerous and diverse of all the dinosaurs in the Cretaceous period. They included the ornithopods, such as the *Iguanodon,* and the duck-billed dinosaurs, like the *Edmontosaurus* and *Maiasaura;* the armored ankylosaurs, including the *Ankylosaurus,* with its protective plating and tail-club; the thick-headed pachycephalosaurs, thought to engage in head butting, such as the *Stegoceras*; and the ceratopsians, four-legged animals with long, bony frills and horns, like the *Triceratops*.

Which **major Cretaceous period dinosaurs** were **herbivores and carnivores**?

All of the remaining sauropods, such as *Saltasaurus, Alamosaurus,* and *Argentinosaurus,* were plant eaters. The most numerous and diverse herbivores in the Cretaceous, however, were the ornithischians, including the duck-billed ornithopods, the horned ceratopsians, the thick-headed pachycephalosaurs, and the armored ankylosaurs. The major carnivorous dinosaurs of the Cretaceous period were from

Raptors—dinosaurs with prominent claws used for hunting, such as the *Utahraptor*—were smaller than the *Tyrannosaurus* but made up for this by hunting in packs (Big Stock Photo).

the saurischian (lizard-hipped) groups, and were theropods. They included the large carnosaurs such as *Giganotosaurus, Carcharodonotosaurus,* and *Tyrannosaurus,* as well as the smaller, more agile dromaeosaurs, such as *Velociraptor, Deinonychus,* and *Utahraptor*. The list is huge, and more fossils of animals are found and added to the list each year.

How did the **distribution** of dinosaur species change during the **Cretaceous period**?

The earlier Triassic and Jurassic periods were characterized by joined landmasses throughout the planet. By the Cretaceous period, these landmasses began to separate, isolating some species of dinosaurs, and leading to different areas having a variety of new species. For example, the titanosaur sauropods were mostly present in former Gondwana, such as South America, while the ceratopsians and hadrosaurs were found mainly in Laurasia. But many of these fossil interpretations are highly debated. One of the main reasons for the present uncertainty about the overall distribution of dinosaurs is the incompleteness of the known dinosaur fossil record and the fragmentation of many dinosaur fossils found within rock layers.

What were some of the **dinosaurs** that lived during the **Cretaceous period**?

The number of dinosaurs that lived during the Cretaceous period was immense—and too many to list here. As more fossils are found, the number will continue to grow. What follows is a partial listing of Cretaceous dinosaurs:

Cretaceous Period Dinosaurs

Name	Common Name	Age (million years ago)	Locality	Maximum Length (feet/meters)
Albertosaurus	Alberta Lizard	76-74	Canada	30/9
Anatosaurus	Rough Tooth	77-73	Canada, USA	43/13
Avimimus	Bird Mimic	about 75	Mongolia	5/1.5
Baryonyx	Heavy Claw	about 124	Europe	28/8.5
Centrosaurus	Horned Lizard	76-74	Canada	16/5
Chasmosaurus	Opening Lizard	76-70	Canada	16/5
Corythosaurus	Helmet Lizard	76-74	Canada, USA	33/10
Deinocheirus	Terrible Hand	70-65	Mongolia	unknown, arms about 10/3

Name	Common Name	Age (million years ago)	Locality	Maximum Length (feet/meters)
Deinonychus	Terrible Claw	121-99	USA	11/3
Dromaeosaurus	Running Lizard	76-72	Canada, USA	6/1.8
Dryptosaurus	Tearing Lizard	74-65	USA	16/5
Edmontosaurus	Edmonton Lizard	71-65	Canada	43/13
Euoplocephalus	Well-armoured Head	85-65	Canada	20/6
Gallimimus	Chicken Mimic	74-70	Mongolia	20/6
Hadrosaurus	Big Lizard	83-74	USA	26/8
Hylaeosaurus	Forest Lizard	150–135	England	20/6
Hypsilophodon	High Ridge Tooth	about 125	England	7.5/2
Iguanodon	Iguana Tooth	130–115	USA, England Belgium, Spain, Germany	33/10
Kritosaurus	Separated Lizard	73	USA; S. Amer.?	30/9.1
Lambeosaurus	Lambe's Lizard	76-74	Canada, USA	50/15
Maiasaura	Good Mother Lizard	80-75	USA	30/9
Orodromeus	Mountain Runner	75	USA	6.5/2
Ouranosaurus	Brave Monitor Lizard	about 110	Niger	23/7
Oviraptor	Egg Thief	80-70	Mongolia	6/2
Pachycephalosaurus	Thick-headed Lizard	about 67	USA	26/8
Pachyrhinosaurus	Thick-nosed Lizard	76-74	N. America	23/7
Parasaurolophus	Near-crested Lizard	76-73	N. America	31/9.5
Parksosaurus	Park's Lizard	70	Canada	10/3
Protoceratops	First Horned Face	85-80	Mongolia	6/2
Psittacosaurus	Parrot Lizard	130–100	Asia	6/2
Rhabdodon	Fluted Tooth	70	Austria, France, Spain, Romania	10/3
Saurolophus	Reptile Crest	70	Canada, Mongolia	40/12
Saurornithoides	Bird-like Lizard	80-74	Mongolia	10/3
Segnosaurus	Slow Lizard	97-88	Mongolia	30/9
Struthiosaurus	Ostrich Lizard	83-75	Austria, Romania	8/2.5
Styracosaurus	Spiked Lizard	76	Canada, USA	18/5.5
Tarbosaurus	Terrifying Lizard	68-65	Mongolia	40/12
Tenontosaurus	Sinew Lizard	125–105	USA	27/8
Triceratops	Three-horned Face	67-65	USA	30/9
Troodon	Wounding Tooth	75-65	Canada, USA	6.5/2
Tyrannosaurus	Tyrant Lizard	68-65	USA	43/13
Velociraptor	Quick Plunderer	84-80	China, Mongolia	6/2

How were **Cretaceous** period **saurischian dinosaurs different** from **Jurassic** dinosaurs?

When compared to the Jurassic period, the Cretaceous period saurischian dinosaurs were still divided into the large, plant-eating sauropods (Sauropoda) and the carnivorous theropods (Theropoda). However, the only important family of sauropods in the Cretaceous were the titanosaurids (Titanosauridae); the other groups, such as the diplodocids and brachiosaurs, lost their dominance. Overall, the theropods became more diverse.

What were the **titanosaurids**?

The titanosaurids (or titanosaurs, the members of the groups Titanosauria and/or Titanosauroidea) were the major group—and truly the only dominant group—of sauropods that lived during the Cretaceous period. They represent a great mystery for paleontologists, too. In fact, only recently have skulls or relatively complete skeletons (such as the *Rapetosaurus*) of any of the roughly 50 species of titanosaurs been discovered.

These dinosaurs had small, slender, pencil-like teeth, similar to the diplodocids. Because of this, until recently, paleontologists put the diplodocids and titanosaurids into the same family—even though parts of both dinosaurs' skulls were very different. In general, the titanosaurs had small, elongated heads, large nostrils, and crests formed from the nostril bones. Most of the titanosaurs lived in the southern continents of Gondwana, especially the southern parts of today's South America and India. Bones of these dinosaurs have also been found in many other places, including Brazil, Malawi, Spain, Madagascar, Laos, Egypt, Romania, France, and the southern United States.

Were there any **other sauropods** that survived into the **Cretaceous period**?

There were very few sauropod species that survived into the Cretaceous period. Most of them, except the titanosaurids in Gondwana, were almost extinct. But there were still a few. For example, although they were abundant in the late Jurassic period, some brachiosaurs survived into the Early Cretaceous, with fossils found in Europe and Africa. There were probably more, but fossil evidence—especially whole skeletons—of Cretaceous sauropods (except for the titanosaurids) is scarce.

Titanosaurids

Saltasaurus: Also seen spelled incorrectly as *Saltosaurus,* this relatively small—about 39 feet (12 meters) long—sauropod had bony plates covering its back in a kind of chain-mail body armor. Fossils have been found in Argentina and, more recently, in Uruguay.

Alamosaurus: This dinosaur was up to 53 feet (16 meters) long, with relatively long forelimbs. It may have been one of the last dinosaurs to go extinct.

Spinosaurus, a species of tetanuran, was a carnosaur considered to be more closely related to birds than almost any other dinosaur species (iStock).

Argentinosaurus: A good candidate for one of the largest land animals that ever lived. This sauropod evolved in South America during the Middle Cretaceous, around 100 million years ago, and may have been more than 85 feet (26 meters) in length.

How did the **theropods** change during the **Cretaceous period**?

During the Cretaceous period, the theropods became much more diverse, with ceratosaurs and the tetanurans well represented. The subgroups of the tetanurans (carnosaurs and coelurosaurs) evolved new species. For instance, the coelurosaurs, composed of the ornithomimosaurs and maniraptorans, became very diverse, although some, such as the carnosaurs, became less dominant.

Did any **ceratosaurs** live in the **Cretaceous** period?

There were some ceratosaurs (Ceratosauria, or "horned reptiles")—theropods that arose during the Triassic period and grew in number during the Jurassic period—in the Cretaceous period. One is from a group of unusual ceratosaurs, called the Abelisauridae, a family (or clade) of ceratosaurian theropod dinosaurs. The Abelisaurids thrived during the Cretaceous period on the ancient southern supercontinent of Gondwana (now South America). One example is the up to 30-foot- (9-meter-) long, strange-looking, horned dinosaur called a *Carnotaurus* ("meat bull").

What were the **carnosaurs** like in the **Cretaceous** period?

The carnosaurs seemed to grow larger in the Cretaceous period. For example, the carcharodontosaurid dinosaurs *Giganotosaurus* ("giant southern lizard") from South America and *Carcharodontosaurus* ("jagged teeth lizard") from North Africa and Morocco were two huge carnivorous theropods. They were probably heavier than the well-known *Tyrannosaurus rex*. In fact, some scientists believe the fossil skull of a *Carcharodontosaurus* found in Africa in 1995 indicates that the dinosaur may have been the largest meat-eater of all—larger than a *T. rex* at 36 to 44 feet (11.1 to 13.1 meters), though this is still debated. Another Cretaceous carnosaur was the *Spinosaurus,* from the Late Cretaceous period, in Niger and Egypt, Africa. One specimen had one of the longest skulls of any carnivorous dinosaur known, estimated to be about 5.75 feet (1.75 meters) long. These dinosaurs are known for the tall spines on their back and lived 100 to 93 million years ago.

What were the **coelurosaurs** like in the **Cretaceous** period?

During the Cretaceous period, there was little change in the coelurosaurs (the clade containing all theropod dinosaurs more closely related to birds than to carnosaurs), except that they continued to diversify. They were composed of ornithomimosaurs, maniraptorans, and tyrannosaurs. In fact, the Cretaceous period probably had the most diverse number of—and biggest—coelurosaurs of any other period.

What were the **ornithomimosaurs** like in the **Cretaceous** period?

The ornithomimosaurs, or ostrich-like dinosaurs, were similar to those in the Jurassic period. They had long, flexible necks and a small head; they also had long forelimbs. These dinosaurs had no upper teeth, and their lower teeth were not well developed. Ornithomimosaurs probably reached lengths of up to 20 feet (6 meters).

What were the **maniraptorans** like in the **Cretaceous** period?

The maniraptorans greatly diversified during the Cretaceous period. They included the major subgroups Aves (what we now call birds), Deinonychosauria, Oviraptorosauria, and Therizinosauria.

What were the **unique characteristics** of the **Deinonychosauria**?

The Deinonychosauria carnivores are known for their switchblade-like second toes. This clade is divided into two further, more familiar clades: the Dromaeosauridae and Troodontidae. The dromaeosaurs (or dromaeosaurids) are dinosaurs the general public associates with the term "raptor." These dinosaurs were members of the maniraptorans, and therefore may share a common ancestor with birds. Some paleontologists even speculate that birds are highly evolved dromaeosaurs.

This group ranged widely in size, starting at just about the size of a large dog and up to 30 feet (9 meters) long. These dinosaurs had large, clawed, grasping hands; a stiffened tail that may have acted as a stabilizer; muscular, toothy jaws; and

What two carnosaurs skeletons were infamously destroyed during World War II?

In 1944, two carnosaurs were among several dinosaurs destroyed by allied bombers at the Bavarian State Museum in Munich, Germany: a *Spinosaurus* and *Carcharodontosaurus*. Other dinosaur skeletons destroyed were types of *Aegyptosaurus* and *Bahariasaurus*.

a large, retractable, slashing claw on the second toe of each hind foot. They were probably agile, predatory dinosaurs. The discovery of their remains revolutionized the way we view dinosaur metabolism and behavior.

The troodontids are a small group with only a few fossil remains found, including several incomplete specimens from North America and Mongolia. These dinosaurs were about the size of a small adult human. They had long skulls; unique recurved and saw-edged teeth; large, flexible hands; and long, slender legs. The troodontids were probably fast, agile hunters. Although they had an enlarged claw on their back feet, it was not as large as those of the dromaeosaurs, and was probably not used in the same way.

The most unique characteristic of the troodontids was the size of their brain case: it was the largest, relative to body size, of all the dinosaurs. This may indicate that the troodontids were the most intelligent of all the dinosaurs, although this is only speculation. The skull had large eye openings (orbits), and there appears to be evidence for well-developed brain centers for sight and hearing. In fact, the ears of troodontids were also unusual among theropods, having extremely enlarged middle ear cavities, possibly indicating acute hearing ability.

What were the **unique characteristics** of the **Therizinosauria**?

The Therizinosauria (or therizinosaurs, also known by the older name, segnosaurs) had so many unique characteristics that they were first grouped with the sauropods. They were then made into a separate group of the saurischians; most recently, scientists placed them together with the maniraptorans, based on various features of their skulls, pelvis, and forelimbs.

Therizinosaurs were large, with heavy builds; they had four toes, similar to sauropods, and were quadrupeds. They also had the hollow bones of the theropods, a pubis bone that pointed backwards, and a relatively short tail. One of their major features was the claws on each hand. In several specimens, the enormous claws measure three-feet (one-meter) long.

More noticeably, the therizinosaurs were very unique. They had large, leaf-like teeth; the skull shows evidence of cheeks, which means they probably ate plants, though this is still highly debated. If so, they would also be one of the only theropods that didn't eat meat. In addition, several specimens seem to indicate that some therizinosaurs were covered with primitive feathers.

Because not many therizinosaur fossils have been found, little is known about their behaviors. Some scientists suggest that these animals were amphibious fish-eaters; others believe these dinosaurs were herbivores, based on their teeth, possible cheeks, and snout. It is also thought that the therizinosaurs were slow-moving animals.

Possibly one of the most well known, if not the most well known, the *Tyrannosaurus rex* was one of the dominant predators during the Cretaceous period (iStock).

What were the **unique characteristics** of the **Oviraptorosauria**?

The Oviraptorosauria (or oviraptorosaurs, or "egg snatchers") were once thought to be ornithomimosaurs, but are now considered to be maniraptorans. The oviraptorosaurs were human-sized, with distinctive skull features, grasping hands, and slender limbs. The skull is the most unique feature of these animals, with a prominent crest containing a large nasal cavity. The function of this crest is unknown, but it may have been used for heat regulation or making sounds. The short, deep skull also had toothless jaws, but they were well-muscled for crushing food. Originally thought to have eaten eggs, the oviraptorids are now thought to have used their muscled jaws to crush mollusk, or perhaps they were omnivorous, eating small animals and some plants.

What were the **tyrannosaurs**?

The tyrannosaurs (Tyrannosauridae, meaning "tyrant lizards," is considered a family of coelurosaurian theropod dinosaurs) were extremely large, carnivorous dinosaurs that lived during the Cretaceous. These dinosaurs, including the well-known *Tyrannosaurus rex,* were the dominant predators of the time, growing close to 50 feet (15 meters) long. They were characterized by long, muscular tails, tiny arms, beady eyes, short, deep jaws, and long legs.

At one time, the tyrannosaurs were placed in the carnosaur group. After all, this group included the large, bipedal, carnivorous dinosaurs, and the *Tyrannosaurus rex* certainly fit all those criteria. However, modern cladistic analysis has shown that the tyrannosaurs are more closely related to the coelurosaurs than the carnosaurs, so they are now placed in the former group.

What are some **examples** of **theropods** in the **Cretaceous** period?

The following are some examples of the various theropod dinosaurs of the Cretaceous period:

Alxasaurus: A therizinosaur from Mongolia, the *Alxasaurus* had longer finger bones than the *Therizinosaurus* (see below). It is thought to be the most primitive-known member of the therizinosaurs, even though it still had the body shape—long neck, short tail, and long hand claws—of the later therizinosauroids.

Caenagnathus: This oviraptorosaur, meaning "recent jawless," grew up to 9.5 feet (2.9 meters) in length and lived about 80 million years ago. Like all oviraptorosaurs, it had a toothless jaw that was well-muscled and perfect for crushing; it also probably had feathers.

Carnotaurus: This bizarre dinosaur from Argentina grew up to 30 feet (9 meters) long; it had stumpy arms and a short head with two horns above the eyes.

Daspletosaurus: This genus of tyrannosaurid theropod dinosaur means "frightful lizard." It is from Canada, and was slightly smaller than the *Tyrannosaurus rex,* measuring 26 to 30 feet (8 to 9 meters) in length. Its skull was huge, reaching up to 3.3 feet (1 meter) in length.

Deinocheirus: Only a pair of 10-foot- (3-meter-) long forelegs and hands have been found of this dinosaur, which may have been one of the largest of the ornithomimosaurs. Fossils of one animal's forelegs are on display at the American Museum of Natural History in New York.

Deinonychus: This dinosaur measured up to 11 feet (3.4 meters) long and weighed approximately 161 pounds (73 kilograms), about the size of a mountain lion. These dromaeosaurs had large sickle-shaped claws on the feet, and some fossil evidence indicates that they had a pack-hunting behavior. Most of the fossils of the *Deinonychus* have been found in North America.

Erlikosaurus: This therizinosaur from Mongolia was long, measuring around 20 feet (6 meters) in length. It also had elongated external nasal openings, slender claws, and a toothless beak.

Gallimimus: This dinosaur (meaning "chicken mimic") probably ate insects and small animals. It lived in Mongolia about 70 million years ago and is the largest and most completely known of the ornithomimosaurs, measuring around 13 to 20 feet (4 to 6 meters) long.

Nanotyrannus: This dinosaur of the genus tyrannosauid, meaning "dwarf tyrant," looks like a small version of a *Tyrannosaurus*. But the fossils are highly debated—are they actually an adult or juvenile tyrannosaurid dinosaur?

Oviraptor: The "egg snatcher or thief" had a bizarre head crest. It was originally thought to prey on others' eggs, but more recent findings show them brooding eggs in nests, although it is still debated as to whether or not they ate eggs. It also was one of the most bird-like of the non-avian dinosaurs. In particular, its rib cage had several features that are typical of birds, especially something on each rib that would have kept the rib cage rigid.

Saurornithoides: This dinosaur is from a genus of troodontid maniraptoran dinosaur from Mongolia, meaning "bird-like reptile." The carnivorous dinosaur measured about 6.5 feet (2 meters) long and had a skull with a long, rather bird-like narrow muzzle. The teeth were small, with many teeth in the upper jaw; the teeth's back edges were serrated. It also had a large brain and large saucer-like eyes, probably for hunting small animals at dusk.

Spinosaurus: This dinosaur was 52 to 59 feet (16 to 18 meters) long, and may have been the largest known carnivorous dinosaur. It had six-foot- (two-meter-) long spines on its back that are thought to have supported a sail; this structure may have played a part in the dinosaur's thermo-regulation. Most of the fossil remains of this dinosaur have been found in North Africa.

Therizinosaurus: This therizinosaur—possibly herbivorous and maybe one of the last and largest of this unique group—had a relatively short tail and huge forelimbs with enormous sickle-shaped claws. They are thought to be closely related to birds (many illustrations depict the *Therizinosaurus* with feathers) and may have been one of the largest theropods, measuring 32 feet (9.2 meters) in length.

Troodon: This troodontid, found in North America, had large eyes and possible binocular vision. It was small, at around 6.5 feet (2 meters) in length, and was only about 3 feet (1 meter) tall. Its long, slender limbs indicate that it was fast on its feet.

Tyrannosaurus: One of the largest and most famous of the land-dwelling carnivores, the *Tyrannosaurus* reached up to 46 feet (14 meters) long and 18.5 feet (5.6 meters) tall. The fossils suggest the females were even larger than the males. One tooth recently took the size prize: it measured 12 inches (30 centimeters) long, including the root when the animal was alive, making it the largest tooth of any carnivorous dinosaur known.

Utahraptor: This dinosaur was up to 21 feet (6.5 meters) long, with a huge, sickle-shaped claw on the second toe that could grow to 9.1 inches (23 centimeters) in length. It is the largest known dromaeosaur found in North America and lived between 132 and 119 million years ago.

Velociraptor: This dinosaur name means "quick plunderer" or "swift seizer". It was a dromaeosaur about the size of a large dog, almost 6 feet (1.8 meters) long, with a weight of approximately 100 pounds (45 kilograms), and a sickle-shaped

A pack of troodons attacks a *Euoplocephal*, a relative of the ankylosaur. While relatively small, these predators were quick and agile hunters (Big Stock Photo).

slashing claw on each foot. Fossils of this dinosaur are found in Mongolia. The *Velociraptor* was also prominently featured in the movie *Jurassic Park* as human-sized terrors; but the actual dinosaur was much smaller and probably had feathers.

ORNITHISCHIAN DINOSAURS

How are the **ornithischian dinosaurs** of the **Cretaceous** period **classified**?

The ornithischian dinosaurs were the most numerous and diverse of the Cretaceous period dinosaurs. Similar to the ornithischian dinosaurs of the Jurassic period, these dinosaurs were all members of the clade called the Genasauria: beaked, herbivorous dinosaurs that were the prey for carnivorous theropods. In one frequently used classification, the Genasauria are divided into the Neornithischia and Thyreophora. The Thyreophora are further divided into the Stegosauria and the Ankylosauria (although in some classifications, these two divisions are combined into one group; in another classification, Eurypoda is added with the Stegosauria and Ankylosauria); the Neornithischia are divided into several types of dinosaurs, the major one being the Ceropoda, which include the Ornithopoda and the Marginocephalia.

What **changes** occurred in the **Thyreophora** during the **Cretaceous** period?

There were many changes in the Thyreophora subgroups of the stegosaurs and ankylosaurs. In particular, few species of the stegosaurs survived into the Cretaceous period. The ankylosaurs fared much better, with the best-preserved fossils coming from Mongolia and China; others are also known from North America, Europe, and Australia. In fact, these plant-eating animals were so huge and heavily armored by the end of the Cretaceous period that they probably didn't have to worry about predators at all.

What **changes** occurred in the **cerapods** during the **Cretaceous** period?

There were more major changes in the cerapods during the Cretaceous period than in the Thyreophora. In particular, the ornithopods became more diversified and the marginocephalians evolved even more over time.

Overall, the Cretaceous period ornithopods still carried the characteristics that made them ornithopods, but they did increase in number and diversify. They were still medium to large plant eaters and, in general, mostly bipedal; they ranged from less than 3 feet (1 meter) tall and six feet (two meters) long to as big as about 23 feet (7 meters) tall and 66 feet (20 meters) long. They further developed their teeth, cheeks, and ability to chew over the millions of years of the Cretaceous.

The animals also seemed to change certain physical characteristics. In particular, the ornithopods could walk, or in some cases trot, on all four feet; at higher speeds, they were probably mostly bipedal. Their feet also started to change, with some evolving into more hoof-like shapes. Others developed hands that could probably grasp vegetation. As they grew in size, some changed their structures to support more weight, including the number of back vertebrae connecting the pelvis to the backbone.

In the Cretaceous period, the ornithopods included such dinosaurs as the hypsilophodonts, heterodontosaurids, iguanodontids, and hadrosaurs. This list includes one of the earliest discovered dinosaurs, the *Iguanodon,* as well as the famous crested and "duck-billed" hadrosaurs, which were the most diverse and successful groups under the ornithopods.

Why did the plant-eating **sauropods die out** while the **ornithopods thrived** during the Cretaceous period?

The majority of the plant-eating sauropods died out in most areas by the end of the Jurassic period and beginning of the Cretaceous period. Some scientists believe this was the result of the switch from eating the usual plants to the new angiosperms (flowering plants). The Cretaceous period ornithopods had teeth that were apparently better adapted to chewing the new plants than the sauropods—especially the ornithopods' multiple rows of teeth inside their jaws. When the teeth in the top row were ground down, new ones shifted to replace the worn ones.

What were the **heterodontosaurs** like?

The heterodontosaurs were fast-moving herbivorous dinosaurs that averaged about 3 feet (1 meter) long. They had canine-like teeth, relatively long arms, and large hands. They were also some of the most primitive ornithopods. Some scientists

An *Iguanodon* is attacked by a group of of *Deinonychus*. Iguanodons were strong and fast herbivores that may have fended off such attacks by using its spikey thumbs (Big Stock Photo).

believe that the animals may have been the first ornithopods with teeth; they had long fangs used for cutting and slicing vegetation or for defense. The animals' hands also may have had some grasping abilities, and they may have used their hands to dig burrows that could have been used as shelter or as estivation "dens" (estivation is lying dormant during the hotter times of the year, which is the opposite of hibernation some animals experience in the winter). Some heterodontosaurs also may have replaced all their teeth at once, depending on the amount of wear.

What were the **hypsilophodontids** like?

The hypsilophodontids were a family of dinosaurs that arose in the Middle Jurassic and are considered one of the first ornithopod groups to appear worldwide. They were a bit longer than the heterodontosaurs, averaging five feet (almost two meters) long, and some scientists compare this animal's size and movements to today's gazelles. They had chisel-shaped cheek teeth overlapping each other, heavy hind legs (most likely for stability when running), and a light build. They may have also nested at the same roosting place each year.

No complete skeletons have ever been found, but a recent discovery of such an animal in Texas rock layers may add new information. One fossil did show something strange: a *Hypsilophodon* with a broken leg that apparently healed. This could mean the animal was able to survive a severe injury; or, as some scientists speculate, it may indicate that other *Hypsilophodons* tended the injured member until it healed.

What were the **iguanodons** like?

The first dinosaur ever discovered by paleontologist Gideon Mantell (1790–1852) was an *Iguanodon*. The iguanodons evolved in the Early Cretaceous period, were larger than the heterodontosaurs and hypsilophodontids, and were fast and strong. They had long, heavy forelegs, and probably walked on all fours. They measured about 33 feet (10 meters) long, had up to 29 teeth per tooth row (on the sides of their jaws for chewing)—and a unique, conical thumb "spike" on the first digit of their hands that may have been used for defense.

What were some **general characteristics** of the **hadrosaurs**?

The hadrosaurs, or "duck-billed" dinosaurs, were some of the most peculiar dinosaurs that thrived during the Cretaceous period, as well as the last ornithopod group to ever appear. Amazingly, these dinosaurs were similar to modern ducks: they had beaks, webbed feet, and a pelvis like a duck. Hadrosaurs also had stiff tails supported by strong, bone-like tendons, and their lost teeth were rapidly replaced. These dinosaurs probably spent most of their lives close to bodies of water, feeding on tough plants; they apparently bore their young on higher ground in large nesting areas.

Most of the duck-billed dinosaurs are found in Late Cretaceous rocks in Europe, Asia, and North America. They are close relatives—and some say descendants—of the earlier iguanodontid dinosaurs. The two subfamilies of the hadrosaurs were the Lambeosaurinae and the Hadrosaurinae.

How did the **Lambeosaurinae** and **Hadrosaurinae differ**?

Several of the hadrosaur dinosaurs are noted for the spacious and bizarre-shaped sinus regions in their skulls. In particular, the Lambeosaurinae had a crest on the skull; the Hadrosaurinae, such as the *Maiasaura* and *Edmontosaurus,* lacked a crest.

The crest on the lambeosaur's skull contained nasal passages looping through the crest and into some very large chambers before they went into the airway. There have been many suggestions as to the reason for such a crest, including its use as a snorkel (but it had no opening to the outside) or as a way to warm the air they breathed (but the climate was already warm). But the most accepted idea is that the crest acted like a resonance chamber, allowing the animals to make deep, loud calls to attract mates (either by the noise, by the odd shapes, or both), scare away predators, or keep a herd of young together.

What were the **marginocephalians**?

The marginocephalians (meaning "fringed headed") were actually latecomers in dinosaur evolution, first appearing in the Early Cretaceous, and chiefly in North America and Central Asia. They are divided into the pachycephalosaurs (Pachycephalosauria, or "thick-headed reptiles") and the ceratopsians (Ceratopsia or "horned dinosaurs").

Why did some ceratopsians have head frills?

The ceratopsian frills may have been the animal's armor to protect itself from predators such as the *Tyrannosaurs rex,* which lived at the same time and locale as the *Triceratops*. There are other ceratopsians that had smaller frills or frills with large openings, which would have been less efficient for defense. Thus, some scientists suggest the frills may have been used as heat radiators, signaling devices, or for attracting mates. In fact, recent studies of the inside of the bony frill indicate that certain parts held different temperatures, thus supporting the idea that it served as a heat radiator.

The marginocephalians are considered a clade of ornithischian dinosaurs. They were closely related to the ornithopods; some scientists believe the marginocephalians may have originated from the ornithopods, especially the heterodontosaurs. The marginocephalians were herbivorous dinosaurs that had a slight shelf or frill at the back of their skull. The frill, or shelf, differed in the two main subgroups of the marginocephalians: the "bone-headed" pachycephalosaurs and the frilled ceratopsians.

What were the **pachycephalosaurs** like?

The pachycephalosaurs, or "thick-headed reptiles," were bipedal dinosaurs with short forelimbs. Some scientists believe the animals' thick skulls were used for head butting: the angle of the skull and backbone indicates that the animal's normal posture was head down with the dome of the head forward.

What were the **ceratopsians** like?

The ceratopsians ("horny faces") appeared in the Early Cretaceous; by the Late Cretaceous, about 100 million years ago, the animals began to diversify in North America and Asia. They were the biggest of all the dinosaur families and lasted a total of about 35 million years. There were frill- and hornless ceratopsians, including the *Protoceratops* from Mongolia, and the unusual, bipedal *Psittacosaurus*. The huge, horned, frilled ceratopsians were found only in the Late Cretaceous period of North America—and were some of the last known dinosaurs to have roamed the planet.

Some ceratopsians were bipedal; some were quadrupeds. It is thought that the quadrupedal ceratopsians evolved from bipedal dinosaurs, as their front legs are shorter than their rear legs. They evolved very strong front legs that were powerful enough to hold up their massive heads, and they ranged from turkey- to elephant-sized animals.

The ceratopsians probably traveled in herds. One reason for this suggestion is that huge fossil remains from hundreds of individual, same-species ceratopsians have been discovered in the western United States in so-called "bone beds." Such a

The *Triceratops*—named after the three horns on its face—was a ceratopsian, a kind of dinosaur distinguished by its prominent head frill (iStock).

large number of the animals in one place suggests that the dinosaurs were traveling together when a tragedy hit, killing them all. Traveling in such large groups, the larger animals, with their intimidating, huge horns and frills, could circle and protect the weaker and younger ceratopsians. Other times, the group could stampede to escape or attack predators.

What were the **general characteristics** of the ***Triceratops***?

The *Triceratops* ("three-horned face") was a ceratopsian that measured up to 30 feet (9 meters) long. It had beaked jaws and three large horns (two long and one short). It also had huge, heavy frills around its long head, but they still don't hold the record for the largest frills, just the most lasting. The *Triceratops*' bones are some of the strongest, most solid dinosaur bones known; they are so well-built that many have survived over 65 million years. These animals also became well adapted to feed on tough vegetation, possessing beaks that could slice up the plants, and rows of teeth to chew.

What were some **ornithischian** dinosaurs during the **Cretaceous** period?

The ornithischian dinosaurs were even more diverse and numerous than the saurischian dinosaurs during the Cretaceous period. The following are some examples of the types of dinosaurs present in this group:

Ankylosaurus: These dinosaurs grew up to 33 feet (10 meters) long. They had a wide head, with triangular horns and bony plates covering their bodies, and for defense they used their stiffened, club-shaped tail. Most *Ankylosaurus* fossils are found in North America.

Tenontosaurus: This ornithopod's fossils are found mostly in western North America. It appears to have been a transitional form to the more advanced iguanodontids—a primitive iguanodont. It measured up to 22 to 27 feet (6.5 to 8 meters) long, had a very long, stiffened tail, and probably walked on all four feet.

Iguanodon: Another ornithopod dinosaur, the *Iguanodon* roamed Europe and North America. It was also the first dinosaur fossil ever found. The *Iguanodon* measured up to about 33 feet (10 meters) long, probably chiefly moved on all fours, and had a conical thumb spike on the first digit of its hand.

Shantungosaurus: Fossils of this largest known hadrosaur were found in China. It measured approximately 50 feet (15 meters) long—as big as many sauropods—and had a skull about 5.35 feet (1.63 meters) long. Its beak was toothless, but its jaws were packed with around 1,500 tiny chewing teeth.

Parasaurolophus: This dinosaur's head crest is the shape of an elongated tube that extends backward behind the skull; many fossils are from North America.

Orodromeus: This North American dinosaur was found in Montana. It grew up to six feet (two meters) in length.

Hypsilophodon: The "high-ridged tooth" dinosaur may have been one of the fastest running ornithischian dinosaurs. It also had a horny beak to cut vegetation and its teeth could easily grind plants.

Heterodontosaurus: This "different toothed lizard" averaged about 3 feet (1 meter) in length; some scientists believe it may have burrowed under the ground during the summer.

Pachycephalosaurus This dinosaur's thickened skull was ornamented with bony knobs, and is thought to have been used in head butting. Most fossils come from North America.

Stegoceras: This dinosaur was about 6 feet (2 meters) long, with the females and males exhibiting different thicknesses in skulls.

Psittacosaurus: Found in Mongolia, scientists believe this "parrot lizard" was a primitive ceratopsian. It's actually called a psittacosaurid ceratopsian dinosaur. In fact, it is one of the most completely known dinosaur genera, with over 400 individual fossils collected so far. The dinosaur was bipedal, with a very rudimentary frill, and is therefore often considered to be frill-less.

Protoceratops: This Mongolian dinosaur was quadrupedal, with a prominent, but short frill. It is therefore often considered to be frill-less. It also lacked the horns seen in later ceratopsian dinosaurs.

Triceratops: The *Triceratops* is another well-known dinosaur from North America. It had a nose horn and paired horns over the eyes. It grew up to 26 feet (8 meters) long, with a short and solid frill; it was one of the last species of dinosaur to roam Earth.

The *Parasaurolophus* sported a weird head crest. Paleontologists are not sure what it was used for, but one possible explanation is that it helped the animal's vocalizations to have increased volume and resonance (Big Stock Photo).

Torosaurus: This North American dinosaur had a longer frill than *Triceratops*. Its six-foot (two-meter) skull is one of the longest known of any land animal.

GENERAL CRETACEOUS DINOSAUR FACTS

What were the **smallest and largest dinosaurs** known in the **Cretaceous period**?

It is difficult to determine the smallest dinosaurs known from the Cretaceous period. In general, the smallest herbivorous and carnivorous dinosaurs could be as small as a chicken. Most of the carnivorous ones ate insects as their main supply of food; and, of course, the smallest herbivores ate plants.

One contender for the smallest is the *Microraptor* (meaning "little plunderer"), a bird-like, crow-sized dinosaur from China. This coelurosaurid theropod was about 16 inches (40 centimeters) long. It may have lived in trees, as its feet were perfect for climbing. Another possibility is the *Wannanosaurus* (named after the Chinese province in which the incomplete skeleton was found). This tiny homalocephalid dinosaur, measuring about 2 feet (60 centimeters) long, was a very primitive pachycephalosaur (related to *Pachycephalosaurus* and *Stegoceras*).

Scientists probably have yet to uncover the largest dinosaurs of the Cretaceous period. One of the largest and most relatively complete skeletons comes from the *Brachiosaurus* of Tanzania, Africa, which measured up to 75 feet (23 meters) in length. But this sauropod didn't make it through the Cretaceous. It evolved around the Late Jurassic period and died out by the Early Cretaceous.

More recent fossil discoveries may lead to even larger sauropods. One includes a massive herbivorous dinosaur *Argentinosaurus huinculensis,* a South American sauropod of the Titanosauridae family, measuring between 130 and 140 feet (40 and 42 meters) long. Another contender is the second-largest sauropod so far found, the *Paralititan* (meaning "tidal Titan"), a titanosaurid sauropod found in Egypt that lived about 100 million years ago.

One of the largest carnivorous dinosaurs yet discovered is the *Gigantosaurus,* which lived in what is now South America and was well over 40 feet (13 meters) long (Big Stock Photo).

Currently, the largest carnivorous dinosaur from the Cretaceous period seems to be a toss-up between the perennial favorite and several newcomers. The favorite is the theropod *Tyrannosaurus,* found in North America and Asia and measured over 40 feet (12 meters) in length. The other challengers include: The longest meat-eating dinosaur yet discovered, the 44- to 46-foot- (13.5- to 14.3-meter-) long *Giganotosaurus* of South America; the *Spinosaurus* a bipedal carnivore that measured about 52 to 59 feet long (16 to 18 meters) from Africa; and the 26- to 44-feet- (8- to 14-meter-) long *Carcharodontosaurus saharicus* of North Africa. All of these huge, meat-eating theropods are thought to have been around the same size—and maybe even heavier—than the *Tyrannosaurus*.

What was the **top predator** among the **Cretaceous** dinosaurs?

This question is difficult to answer. The carnivorous dinosaurs were all vicious as they attacked their prey. Some, like *Tyrannosaurus rex,* used their size to catch their meals; while others were pack hunters, such as the *Velociraptor,* which trapped prey by working together as a unit. Or was it better to be fast or even agile, such as the *Utahraptor* or *Megaraptor*? In other words, the "top predator" is a matter of opinion.

What is the **oldest-known horned dinosaur**?

The oldest-known horned dinosaur is the *Zuniceratops christopheri*. The fossil remains of this animal were recently discovered by Chris Wolfe, an eight-year-old third-grader from Phoenix, Arizona. This dinosaur lived some 90 to 92 million years ago, had three horns, and might have been 10 to 12 feet (3 to 4 meters) long, with a weight of approximately 200 to 250 pounds (100 to 150 kilograms). The remains were discovered in western New Mexico and include jaw parts, the brain case, teeth, a horn, and the brow.

Where did the **last dinosaurs live**?

The last dinosaurs apparently lived in the western regions of North America. Their remains have been found through Late Cretaceous period rocks, while in other regions of the world they disappeared well before the end of this period. Two of the last known species of dinosaurs to arise in the Late Cretaceous period were the *Triceratops,* a herbivore, and *Tyrannosaurus rex,* a carnivore.

Did **any dinosaur species survive** the **extinction** at the end of the **Cretaceous period**?

The general consensus among scientists is that no dinosaur species survived the mass extinction at the end of the Cretaceous period. There are a few paleontologists who don't believe that the dinosaurs died off at the end of the Cretaceous period, but actually lived on, dying out gradually into the Cenozoic era. This theory will continue to be highly debated until dinosaur fossils are proven to exist past the Cretaceous period in rocks. Some scientists claim they have found such evidence, but their findings are still controversial.

In addition, if the definition of dinosaurs includes the birds, then, yes, this family of dinosaurs did survive the extinction. After the Cretaceous period, birds greatly diversified into numerous species, and as we well know, these are species are living everywhere on the planet today.

OTHER LIFE IN THE CRETACEOUS

Besides dinosaurs, what **other animals** were present during the **Cretaceous period**?

Similar to the Jurassic, there were many other animals that surrounded the dinosaurs, competing for space and food. Most of them are familiar from the Triassic and Jurassic periods; while others diversified and evolved. But not all creatures survived; many became extinct, probably because competition increased.

Amphibians

Frogs, salamanders, newts, toads and caecilians, all modern amphibians: They continue to evolve and diversify.

Reptiles

Turtles: Archelon, a large sea turtle, grows up to 12 feet (4 meters) in length.

Snakes: Earliest known snakes appear.

Crocodiles: Many crocodiles become massive, including the *Deinosuchus,* a large terrestrial crocodile that reached 50 feet (15 meters) in length.

Lizards: True lizards continue to evolve and diversify.

Mammals

Triconodonts: Late Triassic to Late Cretaceous mammals; one of the oldest fossil mammals; three cusps of teeth in a straight row give them their name.

Symmetrodonts: Late Jurassic to Early Cretaceous mammals; they had upper and lower cheek teeth with many cusps in a triangular pattern.

Multituberculates: Late Jurassic to Late Eocene mammals; they had cheek teeth with many cusps in more than one row; they probably filled a "rodent" niche that had once been filled by cynodont therasids, and was later filled by true rodents.

Monotremes: First appearance, Early Cretaceous to the present; these animals eventually led to the true mammals (especially with hair and mammary glands). Today, there are three species of monotremes, which are mammals that lay eggs: the Australian duck-billed platypus and two kinds of echidnas (spiny anteaters), which live in New Zealand and Australia.

Early marsupials: First appear during the Middle Cretaceous and survive to the present; these pouched animals had distinct lower and upper cheek teeth.

Early placentals: Middle Cretaceous to the present; the mother nourished her developing fetus through a placenta; the cheek teeth were even more elaborate; the Late Cretaceous placental mammals include insectivores, or mammals that ate insects.

Were there any **flying animals** during the **Cretaceous period**?

The flying animals during the Cretaceous period were the pterosaurs, birds, and various winged insects. There were over 50 species of pterosaurs, and they were found everywhere except Antarctica. They all disappeared at the end of the Cretaceous period. The largest flying animal of all was the *Quetzalcoatlus,* a late Cretaceous period pterosaur found in Texas. This giant of the sky, though not a dinosaur, was comparable in size to modern small airplanes, with an estimated wingspan of 36 to 50 feet (11 to 15 meters).

The birds and winged reptiles that emerged during the Jurassic period greatly diversified during the Cretaceous period. There were also species with reduced wings, such as the flightless, ground-dwelling *Patagopteryx* that looked like a

The *Cearadactylus* was a fish-eating pterosaur that lived during the Cretaceous period in what is now the southern part of North America. There were over 50 species of pterosaurs (iStock).

chicken with very short wings; and the aquatic diving bird, *Baptornis,* which also had tiny wings, webbed feet, and sharp teeth.

Winged insects also greatly diversified quickly during the Cretaceous period, probably in response to the arrival of flowering plants.

What were the **major marine animals** living during the **Cretaceous period**?

As with all life in the Cretaceous, there was a mixture of the older groups and the emergence of modern groups. The Mesozoic "marine revolution" occurred during the Cretaceous period, and included the appearance of new, modern predators that could feed on the older, hard-shelled forms.

Many modern families of marine animals appeared during the Cretaceous period, including the modern crabs, clams, and snails; sharks also evolved into their modern families by the late Cretaceous. Larger animals included mollusks and lobsters. Many bony fishes continued to evolve from much earlier periods than the Mesozoic era.

Marine reptiles still lived in the seas—most of them until the end of the Cretaceous. Mosasaurs, or marine lizards with paddle-like flippers that grew up to 33 feet (10 meters) in length, were around until the Late Cretaceous. Plesiosaurs and ichthyosaurs still swam in most of the oceans, but they died out in the Late Cretaceous. There were also a few marine crocodiles left over from the Jurassic period.

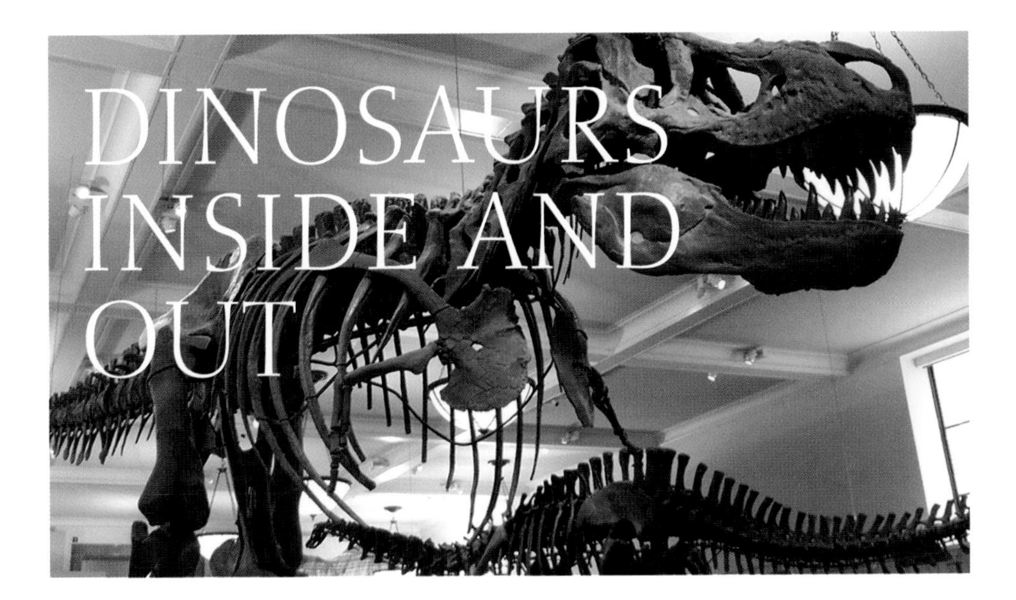

DINOSAURS INSIDE AND OUT

GROWING BONES

Why are **fossil bones** important to our understanding of **dinosaurs**?

The study of fossil bones is extremely important to our understanding of dinosaurs because, in most instances, this is the only way to obtain knowledge about these animals. Other than footprints and rare fossil remains of skin, eggs, and certain (rarely found) petrified internal organs, dinosaur fossilized bones (and teeth) are the most common parts that survive over long spans of time. Depending on the way the bones lie within the rock strata, they provide important clues as to how these animals looked, ate, walked, socialized, lived, and, in some cases, died.

What does the **dinosaur skeleton** tell us about the animals' **soft parts**?

There are several ways in which scientists interpret soft parts based on a dinosaur's skeleton. For example, some muscles leave attachment scars on well-preserved bones. The size of an animal's skull sometimes tells us the relative size of the brain. The holes and hollow spaces in the bones could indicate pathways for nerves. Most importantly, the comparison of dinosaur bones to those of living animals can give us a good idea of the size, anatomy, strength of certain dinosaurs, and sometimes the internal configuration of a dinosaur's organs.

What is the **composition** of most **dinosaur bones**?

Most dinosaur bones, including teeth, are made of calcium phosphate, a material that is very hard and resistant to destruction. This explains why many dinosaur bones survive in rock layers. Those bones that didn't survive were usually destroyed by certain geologic processes. For example, many were crushed by moving rock (from earthquakes or volcanic eruptions), or dissolved by naturally occurring acidic water moving through cracks in the rock.

How does **bone composition** differ between **dinosaurs** and **modern reptiles**?

The bones of a dinosaur and a modern reptile are very similar, with few minor additions or subtractions to bone composition over the millions of years of evolution. Scientists determine this by analyzing specially prepared bone tissues. The tissues are embedded in a synthetic polymer, sectioned, and ground to the right thickness; then they are thinly coated in carbon for analysis in a scanning electron microscope (SEM).

In one instance, the bones of a modern alligator and a *Tyrannosaurus rex* were compared using an SEM with an energy dispersive X-ray (EDX) analyzer. The results showed that both bones had calcium and phosphorous as their major elements, with both elements present in the same ratios in ancient and modern bones. Some trace elements, such as magnesium, aluminum, silicon, and sodium, were also present.

Did **dinosaurs differ** in **bone structure**?

Overall, each dinosaur had a skeleton made up of the same basic structures: the skull, spine, ribs, shoulders, hips, legs, and tail. But individual dinosaur fossil bones do have structural differences. This is apparently dependent on several factors, including where the bones were located in the dinosaur, the bone's purpose or purposes, and the species of dinosaur. In general, dinosaurs that depended on speed needed long, light bones, while larger, slower-moving dinosaurs needed strong, solid bones.

Some of the best examples of the differences between dinosaur bone structures are seen in the bipedal and quadrupedal herbivores. The large, heavy sauropods walked on all fours; they needed strong legs to support their enormous weight, so their bones were huge and solid. The smaller, fast-running bipedal herbivores like *Dryosaurus* needed to be fast; thus, they had long, thin-walled bones. These bones were essentially hollow tubes, and the insides were filled with a light bone marrow. This gave them a strong, flexible, but light-weight, structure, enabling them to move swiftly when circumstances demanded it—such as running from a predator.

When did **primary bones develop** during a dinosaur's life?

Scientists believe primary bones, also called fibro-lamellar bones, were formed during the rapid growth phase of a dinosaur's life—in particular, when the dinosaur was young. These bones were very similar in structure to bones with blood vessels found in birds and mammals; the dinosaur's primary bones also contained blood vessels, which helped them to grow fast. These tissues are especially noticeable in fossil dinosaur leg bones.

What were dinosaur bones made of?

Contrary to popular belief, there is not just one type of dinosaur bone; these complex structures formed the skeletons of complex animals. There are volumes of dinosaur bone tissue studies—most based on the various stages of growth and development—that are too involved to adequately address in just one book.

Overall, there were three main types of dinosaur bone tissues: the primary bone, Haversian (or secondary) bone, and growth ring bone tissue. These varied between the different bones in a dinosaur and sometimes even within individual bones. And they certainly differed from dinosaur to dinosaur.

When did **Haversian** (or secondary) **bone tissue develop** during a dinosaur's life?

In some dinosaurs, the primary bone tissue was later replaced by Haversian bone tissue in a process called remodeling. These tissues had many blood vessels with dense bony rings around them. This type of bone, similar to those of large modern mammals, had more strength and was more resistant to stress.

When did **growth ring bone tissue** develop during a dinosaur's life?

The growth ring bone tissue, found in some dinosaur bones and modern, cold-blooded reptiles, look similar to growth rings found in trees. Tree rings grow each year, responding to changing seasonal conditions. By counting the rings, it is sometimes possible to tell the average age of a tree.

The presence of similar structures in certain dinosaur bones suggests the animals' growth rates slowed down later in life. One interpretation is that the animals became more reptile-like. One major problem is interpreting a dinosaur's growth ring bones: unlike a growth ring on a tree, no one knows the amount of time represented by each growth ring in a dinosaur bone.

What do these **three types** of **bone tissue** indicate about **dinosaur physiology**?

The presence of these three bone tissue structures suggests dinosaurs had a unique physiology lying somewhere in between cold-blooded reptiles and warm-blooded birds and mammals. Perhaps this unique physiology made dinosaurs extremely adaptable, enabling them to dominate the land for around 150 million years.

Where are **extremely large dinosaur bones** being found and why?

Some of most consistently large dinosaur bone discoveries are being made in South America—especially in northwest Patagonia, in Argentina. This includes such

Dinosaur bones are used for more than impressive museum displays. They can tell scientists a lot about how dinosaurs behaved and adapted to their environments (iStock).

examples as the *Argentinosaurus* and *Megaraptor*. Although scientists believe they know why these animals evolved differently from their northern counterparts, they truly don't know why the animals became so large.

What happened to **North and South American dinosaurs** over time?

One theory to explain the large South American animals involves locality. At the beginning of the Mesozoic era, all the land on Earth was merged into the continent of Pangea. During the Jurassic period, the supercontinent broke into the continent of Laurasia (which would eventually become North America and Asia) and Gondwanaland (eventually Africa, Antarctica, Australia, India, and South America). And not long afterward, South America and Africa began to split apart.

Most scientists theorize these splits were the pivotal points in these dinosaurs' evolution. For a short time, the dinosaurs crossed a land bridge from North to South America; geologic activity eventually destroyed the bridge, cutting off access and allowing creatures like the migrated *Megaraptor* to evolve separately in the south. The North American animals, such as the *Tyrannosaurus rex,* developed a specialized skull, forelimbs, and pelvis; the South American dinosaurs, such as the *Giganotosaurus,* maintained most of the general features of their ancestors and became much larger.

Other scientists believe that southern *Megaraptors* and their northern counterparts originally evolved separately from common ancestors. They suggest that the reason why the two carnivorous giants like the *Tyrannosaurus rex* and *Giganotosaurus* resembled each other was possibly due to similar environmental conditions (this is called parallelism); when the landmasses began to break up, the animals continued to evolve separately.

Why did **larger dinosaurs** evolve in **South America**?

One theory to explain the large South American animals involves locality. Scientists are excited about the possibility that environmental conditions differed greatly on all the continents, causing the animals to evolve even larger in South America because of that distinct environment. Just what those certain environmental conditions were is unknown. But if something in the environment truly caused distinct differences in the dinosaurs, there are probably many different types of dinosaurs still to be discovered all around the world. Only time will tell if scientists can determine the true reasons why the South American dinosaurs were giants of the Cretaceous world.

How **many bones** made up the **average dinosaur skeleton**?

Although the largest dinosaurs may have had a few more bones in their necks and tails, the number of bones in the average dinosaur was approximately 200.

In general, how do scientists decide the **bone positions** of **dinosaur skeletons**?

Determining where bones go in a dinosaur skeleton is not an easy task. Scientists have to compare every bone with other dinosaur skeletons, as well as with modern species of reptiles, and hope to find a skeleton in a "death pose" that was close to its living structure. Many times in the past, certain parts of a skeleton were put in the wrong place. For example, heads of certain dinosaurs have been put on the wrong skeleton, and the thumb spike of the *Iguanodon* was first interpreted as a nose spike.

The positions of bones in dinosaur skeletons are determined using what scientists call an "anatomical direction system," and only includes what is internal (in other words, it is not based on external conditions). This system uses pairs of names to determine certain directions based on the average (or standard) posture of tetrapods, with the back up, belly down, head pointing forward, and all four legs on the ground.

Each pair of names denotes opposite directions similar to when we refer to north and south. Here are four examples of such paired names:

Anterior and posterior: The direction of anterior is toward the tip of the snout, while the posterior direction is toward the tip of the tail. This is analogous to front and back, respectively.

Dorsal and ventral: Dorsal means toward and beyond the spine, while ventral means toward and beyond the belly. These are analogous to up and down, respectively.

Why do some dinosaur fossil bones have missing pieces?

For quite some time, paleontologists have noted certain dinosaur fossil bones have missing pieces—for example, a set of teeth without a jaw bone—and others seem pitted and grooved. Recently, scientists may have discovered the answer: ancient insects that gnawed on the dinosaur bones.

The study of the 148-million-year-old remains of a *Camptosaurus,* a plant-eating sauropod, found traces of insects that were probably a form of beetle from the family *Dermestidae.* Scientists speculate that after other insects ate their way through the flesh, horns, or the various soft parts of the dinosaur, the beetles, whose descendents still exist today, munched on the bones.

Medial and lateral: These are directions referenced to an imaginary plane located in the center of the body, running from tail to snout. Medial means closer to this central reference; lateral means farther out.

Proximal and distal: These are normally used to indicate directions in the limbs and sometimes the tail. Proximal means closer to the trunk or base of a limb, while distal means farther out from the trunk or from the base of the limb.

What are the **major parts** of a **dinosaur skeleton**?

A dinosaur skeleton is divided into two major parts: the skull and all the rest of the bones, which are normally referred to as the postcranium (posterior to the cranium). This postcranium can be further divided into the bones of the spine, trunk, and tail (axial skeleton), and the bones of the limbs and limb girdles (appendicular skeleton).

In general, what comprises a **dinosaur's skull**?

The skull of a dinosaur is made up of the teeth and all the bones in the head. These bones can occur in pairs, on opposite sides of the head, or singly (usually around the middle plane of the skull). There are two major sections of the skull. The upper part, or cranium, contains the braincase, nostrils, upper jaw, and eye sockets. The other section is the lower jaw, which is made up of the right and left lower jaws (mandibles).

Examples and locations of dinosaur skull bones include: the cheekbone (jugal), located below the eye opening (orbit); the postorbital, a small bone located behind the eye opening; and the lacrimal, a bone that separates the eye opening and the opening forward of the eye.

There are several interesting details about dinosaur skulls. For example, there are more than 30 bones in the skull of a dinosaur. Most dinosaurs had unusually rigid joints between their skull bones called sutures (similar to sutures in a human skull). There were also kinetic skulls, such as the *Allosaurus fragilis,* in which sev-

We often think of dinosaur fossils being discovered like this one, with just about all the pieces perfectly preserved. In most cases, though, only incomplete skeletons are found, for various reasons (iStock).

eral of the skull bones were joined but could still move, probably so they could stretch parts of the skull in order to gobble down extremely large chunks of meat.

What is the **axial skeleton**?

The axial skeleton is one of the two sections of the entire dinosaur skeleton. It is made up of the trunk, spine, and tail, essentially forming the "foundation" to which the animal's limbs and skull were attached. In other words, it includes the so-called "backbone" of the dinosaur and the ribs. The backbone (or vertebral column) is divided into four segments: the neck (cervical), the back (dorsal), hip (sacral), and the tail (caudal). It included numerous individual bones known as vertebrae. The ribs were long, narrow bones attached to the vertebrae of the cervical and dorsal segments. These were paired bones—one on each side of the backbone—and extended downward to protect the internal organs. They included the neck ribs and the belly ribs (gastralia, or the bones that protected the digestive tract and other internal organs).

What were the **vertebrae** of a **dinosaur**?

Vertebrae were the numerous individual bones that made up the backbone (vertebral column) of the dinosaur. Each individual vertebra was a roughly cylindrical-shaped piece of bone (centrum); on top of the vertebrae were neural arches, triangular arches of bone covering the spinal cord. The spinal cord in a dinosaur would

What is one of the biggest fossil dinosaur bones found to date?

One of the largest fossil dinosaur bones is from the *Futalognkosaurus,* a titanosaurian dinosaur found in Argentina and described in 2007. This new genus of plant-eating dinosaur was more than 100 feet (30 meters) in length from tail to nose and lived about 90 million years ago. No doubt more such huge dinosaurs will be found in the future.

run between the centrum and neural arch. A bony neural spine projected up from the neural arch and was where the back muscles were attached. Some dinosaurs, in addition to these basic features, had very complex vertebrae with all sorts of ridges and projections.

Each segment along the backbone had vertebrae that were specifically shaped to help that segment function. For example, the hip (sacral) vertebrae were fused together in dinosaurs in a structure called a sacrum; this provided support and strength for the hips. The neck (cervical) vertebrae were specifically shaped to provide flexibility, allowing the dinosaur to move its head around freely.

It's impossible to pinpoint how many vertebrae each dinosaur had, as they varied greatly among all groups. In general, the neck bones held 9 to 19 vertebrae, the back had 15 to 17 of these bones, the hips held 3 to 10 vertebrae, and the tail had from 35 to 82 vertebrae, depending on the dinosaur.

How was the **vertebral column** in **ornithischian dinosaurs** strengthened?

The vertebral column in ornithischian dinosaurs was strengthened by means of structures called "ossified tendons." These were actually tissues that connected the vertebrae together. They became filled with calcium, literally turning to bone (called ossifying). The result of this was a stiffening and strengthening of the connection between vertebrae, resulting in a strengthened backbone.

On ornithischian dinosaurs, ossified tendons appear as bony strands that look like spaghetti. Different names are given to them depending on where they are located on the dinosaur. For example, those between the tail vertebrae are called hypaxial tendons. In ornithischian dinosaurs, one purpose of the ossified tendons was to stiffen the base of the tail, making it more rigid with respect to the hips, while the tip of the tail was left mobile and flexible.

How were connections between **theropod dinosaur vertebrae** strengthened?

Some advanced species of theropod dinosaurs had strengthened vertebrae connections, but this was not accomplished by means of ossified tendons, as with the ornithischian dinosaurs. The theropod's prezygapophyses elongated and grew over several vertebrae—sometimes as many as 12 vertebrae—and it strengthened and stiffened the area.

What are some of the **richest dinosaur bone areas** in the world?

Some of the richest dinosaur bone graveyards are found in three diverse regions of the world: the Patagonia region of Argentina, the Gobi Desert in China, and the western United States. Most recently, scientists have found one of the largest plant-eating dinosaurs, the *Argentinosaurus,* and the largest carnivore, the *Gigantosaurus carolinii,* in the Patagonia region of Argentina, which has one of the richest and longest-term fossil records in the Southern Hemisphere. No doubt more dinosaur bones will be discovered in these three unique places in the future.

Most dinosaurs walked on their toes rather than on the soles of their feet, which allowed them to move with greater speed (iStock).

What was **unique** about **dinosaur feet**?

The feet of dinosaurs were very different than those of humans. Because of their unique foot structure, most dinosaurs walked on their toes. There were a number of advantages to toe walking. For example, walking on toes increased a dinosaur's leg length, which in turn increased the animal's stride lengths. This translated into more speed, especially for bipedal dinosaurs such as the theropods and some herbivores like the ornithopods. The advantage was better hunting or ability to escape. This type of motion also saved energy because the body didn't have to be raised and lowered every time the foot was lifted.

The long bones of the feet (metatarsals) were bunched together for strength. They were oriented upwards in a digitigrade position toward the ankle joints at an angle, essentially lifting the metatarsals off the ground. (The term digitigrade literally means "toe-walking.") This meant that only the toes made contact with the ground as the dinosaur walked or ran. In humans, the foot bones are parallel to the

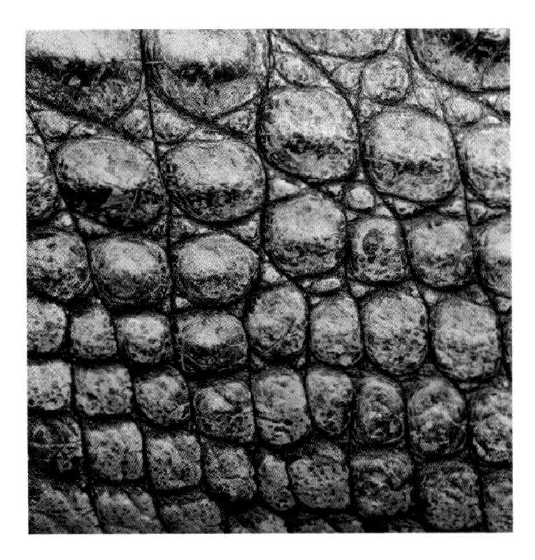

Some dinosaur species had bony structures in their skins (osteoderms). Today, crocodillian species still have osteodermic skin, though the bone embedded in the hide is concealed beneath the surface (iStock).

ground so that the whole foot, not just the toes, makes contact. This is called plantigrade, or "sole-walking."

The toes of most dinosaurs' feet were normally long and slender, which allowed the animals to grip the ground and have better balance. Most of the dinosaurs had only three toes for walking or running. We see the result in former muddy or sandy areas as dinosaur trackways complete with bird-like footprints.

Not every dinosaur had three long and slender toes. There were some exceptions. For example, large quadrupeds, such as the sauropods, had feet more like those of today's elephant, with short and broad toes.

What was **unique** about the **feet** of most **quadruped dinosaurs**?

Unlike the bipedal (two-legged) types, the feet of most quadruped (four-legged) dinosaurs had shorter foot bones and a set of broad, stubby toes. This arrangement of foot bones was very similar to the feet of modern elephants.

Footprints of these types of dinosaurs were short and round, indicating that the bones of the feet were raised off the ground by a thick, fibrous, wedge-shaped heel pad. This heel pad enabled these huge animals to conserve a large amount of energy as they moved around. The ankle did not have to be raised and lowered during walking, a motion that would have lifted the entire body weight up and down. For dinosaurs like the large sauropods, this motion would have meant expending a great deal of energy.

What were **osteoderms**?

Osteoderms were bony growths located on the outside of some dinosaurs' skins, and were anchored in the skin by connective tissue. The most familiar examples of osteoderms were the spikes and plates that made up the armor of stegosaurs and ankylosaurs.

Can you tell the **difference** between **male and female dinosaurs** from their **bones**?

No, it is really not possible (at present) to determine the gender of a dinosaur by looking at its fossil bones. There are few clues to determine a dinosaur's gender, although some scientists believe certain species may have had features that distinguished gender. For example, hadrosaurs (duck-billed dinosaurs) sported certain types of bony head crests, but just which gender *had* the crest is unknown. Currently, scientists are trying to base dinosaur gender ideas on examples from today's animal world, and even then it's still almost impossible to determine a male from a female dinosaur.

What do **dinosaur bones** tell us about a **dinosaur's stance**?

All dinosaur skeletons show that these creatures had a "fully improved stance." In other words, dinosaur legs were held straight under their bodies at all times. This enabled dinosaurs to grow bigger, cover longer distances, and move faster, compared to their reptile relatives that had their legs spread out on either side of their body. The dinosaur stance also enabled some of the animals to become bipedal (walk on two legs). It also helped all dinosaurs with something called "locomotor stamina," which is the ability to run and breathe at the same time.

What can paleontologists tell about the **lifestyle** of a **large herbivore (sauropod)** from its fossil **bones**?

The bones of giant dinosaurs like the *Diplodocus* tell us a great deal about this herbivore. Its legs were thick and widely spaced, acting as pillars to hold up the cross beams of its shoulder bones and hip girdle. The vertebrae across the hip were fused for strength, allowing it to support an almost 11-ton (10-metric-ton) body weight. The legs ended in short, broad feet (similar to an elephant's) with claws on the back foot used as an anti-slip device. The bone structure limited the dinosaur to a normal walking pace of approximately 4 miles (6.4 kilometers) per hour, although they could have moved modestly faster for short periods of time. Thus, they were thought to be large, slow moving, four-legged walkers. In addition, the large adult sauropods were probably relatively immune from predators because of their large size.

What were **unique skeletal features** of **large sauropods**?

There were a number of unique skeletal features of large sauropods that evolved mostly to support their massive size and weight. These dinosaurs were similar in construction to a suspension bridge: the front and rear legs acted as underlying supports for the backbone; in turn, the backbone was supported from above by ligaments and muscles of the back. The vertebrae of the backbone were joined together so the neck, back, and tail were bent slightly upwards at the ends, spreading the massive weight toward the ends.

The top leg bone was shaped to swing underneath the body. The knee was constructed so the leg could move back and forth, similar to the knee of humans. The ankle had very limited movement, with no possible sideways motion. These and other skeletal features helped the large sauropods to support and move their tremendous bulk.

What can paleontologists tell about the **lifestyle** of a **small herbivore** from its fossil bones?

One good example of small herbivores is the *Hypsilophodon,* an ornithischian with a skeleton very different from its larger sauropod cousins. The small dinosaur's

Have scientists ever found soft tissue in dinosaur bones?

Yes, recently, and contrary to what most scientists believed could happen, paleontologists have found what seems to be soft tissue from the femur of a 68-million-year-old *Tyrannosaurus rex*. This is a surprising discovery if it is true—after all, soft parts decay quickly, and organic molecules are thought not to exist after 100,000 years.

Scientists studied the microstructure and organic components of the *T. rex* bone, removing the minerals from the fragment; they found a stretchy bone matrix material left behind that seemed to show blood vessels and other organic features. Because it is thought that living birds are closely related to dinosaurs, the scientists compared the branching vessels to a contemporary ostrich, and the vessels were indeed very similar. But like all new discoveries in science—and especially with dinosaurs—there is more work to do to confirm this finding.

entire structure seemed "shrunk down," giving it strength with minimum weight. Its bones were hollow and thin-walled for lightness, and the thigh bone was very short for rapid stride movements. The small dinosaur's feet were long and thin, with long, upper-foot bones (metatarsals), and it had short, sharp claws for gripping the ground. The long tail bone was stiffened by bony rods, and probably swung from side to side, helping it to quickly change directions. All of these structures paint a picture of a small, two-legged herbivore using its bony features to swiftly run and maneuver as a defense against predators.

What can paleontologists tell about the **lifestyle** of a **large carnivore (theropod)** from its fossil **bones**?

The *Tyrannosaurus* is currently the most recognizable of the large carnivorous dinosaurs. Its skeleton had heavy, large bones, with massive vertebrae, hip girdle, and thigh bones. The upper foot bones (metatarsals) were locked together for strength, while the toes were powerful and short. The knees show evidence of thick cartilage, similar to modern birds.

There are two scenarios concerning the speed and mobility of the *Tyrannosaurus* based on the animal's skeletal structure, and both sides point to the same evidence to bolster their claims. One group thinks that the skeletal structure of a *Tyrannosaurus* caused the animal to move at a slow pace, which limited its main hunting abilities to either scavenging or ambush techniques. The other group suggests that the dinosaur's bone structure, along with the animal's massive musculature, enabled the *Tyrannosaurus* to run and sprint, making it an active, dangerous hunter.

Until more direct evidence is gathered, the most agreed upon theory is based on a major deduction: the *Tyrannosaurus* would need more meat than was available from scavenging, so it would have to hunt. To do so, it would have to at least match the speed of its prey. In other words, it would have to keep up with such dinosaur prey as the herbivores *Triceratops* and *Edmontosaurus* (ornithischians), both of which are thought to have reached 9 to 12 miles (14 to 19 kilometers) per hour for short bursts.

ABNORMAL DINOSAUR BONES

What do **dinosaur bones** indicate to paleontologists about the **health** of these animals?

Dinosaurs, in general, seem to have been relatively healthy animals, if the evidence from—and the interpretation of—the fossil records are to be believed. A few fossilized bones have shown evidence of abnormalities, such as asymmetrical bone growth, healed traumatic and repetitive stress fractures, arthritis, ossification (the development of bony-like material) of spinal ligaments, and the fusion of the animal's spinal bones (or vertebrae).

What caused **asymmetrical bone growth** in dinosaurs?

Asymmetrical bone growth could have occurred, for example, when a tendon was ripped off the bone. This probably happened during some form of exertion, such as running after prey for carnivores, or trying to escape a predator for herbivores. If bone growth continued in this area after the tendon was torn, it would often grow back in an abnormal shape.

What caused **traumatic** and **repetitive stress fractures** in dinosaur bones?

Another feature that has been seen in some dinosaur bones, such as *Tyrannosaurus* and *Iguanodon,* are healed traumatic fractures. Such fractures may have occurred during struggles with other dinosaurs or even mating activity.

Another type of fracture, called a stress fracture, apparently occurred in dinosaurs as a result of repetitive stresses to the bone. Stress fractures found in ceratopsians, such as the *Triceratops,* have often been blamed on foot stamping, sudden accelerations in response to predators, or even fractures brought about during long migrations.

What do **traumatic fractures** in the bones of **large theropods** (carnivores) tell us about the **behavior** of these animals?

Some bones from carnivorous dinosaurs do show signs of traumatic fractures. One example was recently found by scientists using X-rays of the rib bones from an *Allosaurus*. Scientists believe the fractures could have been caused by a belly flop onto the hard ground while running. This suggests that these large theropods were

Scientists used to believe that no DNA could have survived the millions of years since the dinosaurs became extinct, but in 2007 they discovered collagen proteins from within a *T rex* bone. This is an amazing discovery that could prove, among other things, how closely dinosaurs are related to birds (iStock).

not sluggish creatures, but were active hunters, running after prey fast enough to crack ribs if they fell during the pursuit.

Have any **dinosaur DNA** or **proteins survived** 65 million years?

It is thought that most dinosaur bones that are at least 65 million years old have all turned to stone, and so there is no hope of DNA or proteins surviving. But a study in 2007 showed that scientists may have been too hasty when proteins were successfully extracted from a *Tyrannosaurus rex* bone. Using special extraction methods, the scientists found what seems to be collagen, which is one of the major proteins found in our own bones. Why is this exciting to paleontologists? By comparing the dinosaur collagen to contemporary chicken collagen, scientists might be able to answer a question that has plagued them for years: how closely are birds and dinosaurs related?

Do **dinosaur bones** show evidence of **arthritis**, a common affliction in humans?

Yes, some dinosaur bones show signs of certain types of arthritis, especially osteoarthritis and inflammatory arthritis. In humans, osteoarthritis, or degenerative arthritis, is common in the elderly. It is the increased deterioration of cartilage around the bone due to age. Inflammatory arthritis, or gout, in humans usually occurs when crystals of uric acid are deposited in a joint. The excess amounts of uric acid are usually unexplainable, but it has often been tied to dietary excesses.

In the vast majority of cases, dinosaur bones show almost no sign of osteoarthritis, leading some paleontologists to theorize these creatures had highly constrained joints, or bone joints with little rotational movement. Two specimens of *Iguanodon,* however, were found to have evidence of osteoarthritis in the ankle bones, or weight-bearing parts of the body. Because scientists don't know the life spans of dinosaurs, they also don't know whether the arthritis was caused by old age. In addition, two tyrannosaurid dinosaur remains showed evidence of inflammatory arthritis in the hand and toe bones—possibly the result of a rich, red meat diet.

What **bone phenomena** do both **humans** and **dinosaurs** share?

Humans and dinosaurs share a process called "diffuse idiopathic skeletal hyperostosis" (DISH), or when the ossification (when something becomes bonylike) of the spinal ligaments stiffens the spinal area. Although it sounds bad, it is a normal process and not recognized as a disease in either humans or dinosaurs.

Creatures such as ceratopsians, hadrosaurs, iguanodonts, pachycephalosaurs, and some sauropods all show DISH—a stiffening of the dinosaur's tail area that made it easier to hold the tail off the ground. Dinosaurs that used their tails as weapons, such as the stegosaurs, needed them to be flexible like whips, so they do not show evidence of this spinal ligament fusion. The discovery of DISH in dinosaurs dovetails nicely with a newer theory that many dinosaurs did not drag their tails, but rather held them off the ground as a form of counter-balance.

Another bone-related phenomena humans and certain dinosaurs share is vertebral fusion, where the bones of the spine (the vertebrae) actually become joined and ossified together (as opposed to the spinal ligaments in the DISH process). In the adult ceratopsians, such as the *Triceratops,* this fusion was limited to the first three neck (cervical) vertebrae, leading to speculation that this was not a disease, but a developmental adaptation. Stiffening in this area may have evolved to better support the animal's massive skull. Current fossils reveal that smaller, perhaps younger ceratopsians had incomplete fusion in this area; whereas the fusion was complete in larger, presumably older animals.

DINOSAUR SKIN

Are there any **fossils** of **dinosaur skin**?

The rarest types of dinosaur fossils are those showing skin texture—but not the actual skin itself. This is because of the fossilization process: the body of the dinosaur needs to be in a dry environment and some soft parts mummify; then the mummified parts leave an impression on the rock, and this is very rare.

The few fossils of dinosaur skin uncovered to date show that most dinosaur skin was tough and scaly, like modern reptiles. For example, there was the tough, wrinkled skin with bony plates of the Cretaceous period hadrosaur *Edmontosaurus*. Similar to the hadrosaurs, ornithopods had thick, wrinkled skin with

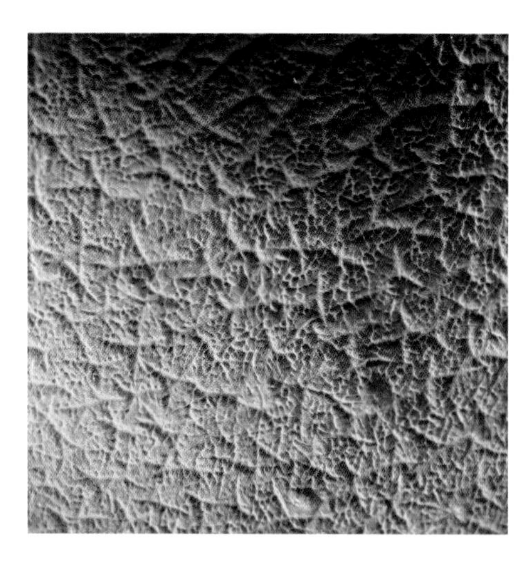

Because it decays quickly, there are very few examples of dinosaur skin that have been fossilized. The impression samples that survive have provided clues, such as the possibility that some dinosaurs had feathers (iStock).

embedded bony knobs of various sizes. Some small theropods, like the recently discovered *Sinosauropteryx,* may have had feather-like features on its skin for heat regulation.

What **dinosaur "skins"** have recently been **found**?

There have been several discoveries of dinosaur "skin" in recent years. One was discovered several years ago by a teenager on his family ranch in North Dakota: a 67 million-year-old hadrosaur (duck-billed dinosaur) with actual fossilized skin, complete with scales and a suggestion of stripes. The dinosaur mummy was found in southwestern North Dakota in 2004; to date, it is one of only about four dinosaurs ever found with fossilized skin. Another one was found in China, complete with a tear caused by a predator. It also shows that below the scales of this *Psittacosaurus* was a thick hide comprised of 25 layers of collagen, which is thought to be similar to modern shark skin.

Did all **dinosaurs** have the **same type of skin**?

The amount of fossilized skin uncovered to date is extremely small, and most of our ideas of dinosaur skin come from extrapolation from modern reptiles. The chances of all dinosaurs having the same skin is probably small, since, over millions of years, these animals adapted to their environment and specialized needs.

Today's reptiles do not have the same types of skin, either. Reptilian skins vary from a lizard's protective covering of scales or plates to the hard bony shell of turtles—all different adaptations to their specific needs and environments.

Have any **dinosaur embryo skin impressions** been found?

Yes, for the first time ever, the skin impressions of dinosaur embryos have been discovered at a large nesting site in the Patagonian badlands of Argentina, South America. The eggs at the site—nicknamed "Auca Mahuevo" ("huevo" meaning egg) by paleontologists because of the large number of dinosaur eggs found there—are approximately 70 to 90 million years old, placing them in the Late Cretaceous period. Some of the eggs contained embryos, with patches of fossilized baby dinosaur skin. The skin had a scaly surface, similar to that of modern lizards. One skin impression had a distinct stripe of larger scales near its center, a section that probably ran down the dinosaur's back.

Have any other **soft parts** of **dinosaurs** been found?

Yes, some soft parts impressions of dinosaurs have been found in the past, but they are rare. Plus, some fossils display the outlines of internal organs; and some animals have fossilized remains inside them of what the dinosaur had just eaten. One of the most exciting discoveries is housed in the North Carolina Museum of Natural Science: the first-ever discovery of traces of a dinosaur heart in a fossil nicknamed "Willo." This dinosaur was a member of the group known as *Tescelosaurus* ("marvelous lizard," living about 66 million years ago), a 600 pound, hog-like plant eater about the size of a pony. Its long, bony tail gave it a total length of about 13 feet (4 meters).

Using special equipment to see "inside" the dinosaur's body, researchers developed a three-dimensional view of the heart. The organ had four highly developed chambers and a single arched aorta. This one discovery—found in North Dakota—changed the way many scientists look at dinosaurs. In particular, it provided evidence that some of the long-extinct beasts were quick and possibly warm-blooded like birds (a complex heart is usually indicative of a high metabolism), not slow and plodding like some contemporary reptiles.

What **dinosaur fossil** found in **Italy** showed the remnants of **soft parts**?

The fossil of a small, baby carnivorous dinosaur—hardly more than a hatchling—was announced in Italy in the 1990s. The dinosaur was actually found by an amateur collector in the southern part of the country in 1981, but he thought it was just the fossil of a bird. In 1993, the fossil collector saw the movie *Jurassic Park* and realized his fossil looked very similar to the movie's *Velociraptor* (in reality, true *Velociraptors* were smaller). In 1998, after the fossil was examined by paleontologists, it was determined to be the bones of a young dinosaur; it was also the first dinosaur ever discovered in Italy.

This dinosaur, called *Scipionyx samniticus,* is about 113 million years old. Although it is a distant cousin of both the *Tyrannosaurus rex* and *Velociraptor,* it is considered to belong to an entirely new family. The fossil also shows something that usually does not survive millions of years of fossilization: soft parts, including a fossilized digestive tract running through the skeleton, from the throat to the base of the tail. Even the wrinkles in the dinosaur's intestines were preserved.

Is there any evidence of **sickness in dinosaurs**?

Yes, there has been a great deal of evidence of dinosaur illnesses. For example, fossils of a dinosaur called *Gilmoreosaurus* showed, in 2003, evidence of a tumor, as well as hemangiomas, metastatic cancer, and osteoblastoma—all major illnesses in organisms. Several other hadrosaurids, including *Brachylophosaurus, Edmontosaurus,* and *Bactrosaurus,* also showed many of these diseases in various studies. Scientists are still speculating on why the dinosaurs had such problems. They could have stemmed from anything from environmental factors to a genetic propensity toward the diseases.

What **color or colors** were the **dinosaurs**?

No one, as yet, has been able to tell anything about the color of a dinosaur's skin. The skin "fades" as it is mummified, and the rocks eventually lend their own color to the fossil. But paleontologists theorize that dinosaurs, like some modern animals, used color and patterns to camouflage and identify themselves. Therefore, dinosaurs' skin colors probably ranged from light and dark browns to greens in various patterns—all earth colors, allowing them to hide or blend in with their environment.

But there may have also been brightly colored, smaller dinosaurs. After all, today's birds—thought to be directly related to the dinosaurs by many paleontologists—are often brightly colored in order to attract a mate or warn other birds (and even predators) away from their territory. Many scientists believe the dinosaur fossils being found with feathers—some with traces of feather pigment—may one day lead to knowledge about dinosaur colors.

There is also some evidence that crocodiles and birds, the closest living relatives of dinosaurs, may have color vision. This suggests that dinosaurs may have responded to colors in their environment, especially bright colors for territorial or mating displays or for identifying prey more easily. It's also interesting to note that crocodiles with color vision are themselves not brightly colored at all.

TEETH AND CLAWS

What **bones** held the **dinosaur's teeth**?

Dinosaurs had two bones in the upper jaw that held teeth. The larger bone holding the majority of the upper teeth were located toward the back of the snout (called the maxilla). The smaller bone holding fewer teeth (called the premaxilla) was located at the front of the skull. Teeth were held by the dentary section of the lower jaw; these teeth are called dentary teeth.

What was the **composition** of **dinosaur teeth**?

Dinosaur teeth were formed from two materials, called dentine and enamel. These two materials were more durable and tougher than bone. Dentine was the softer of the two, and formed the core of each tooth; the outer surface was covered with the harder enamel.

What do we know in general about **dinosaur teeth**?

Scientists know that most dinosaurs had more teeth than humans. In addition, the dinosaurs would shed their teeth throughout their life, much like animals such as sharks do today. For example, the hadrosaurs had hundreds of teeth waiting to replace their worn out teeth. Other dinosaurs had no teeth at all, but had a beak similar to a bird, such as the ornithomimids. Some dinosaurs also had a combination of a beak and teeth.

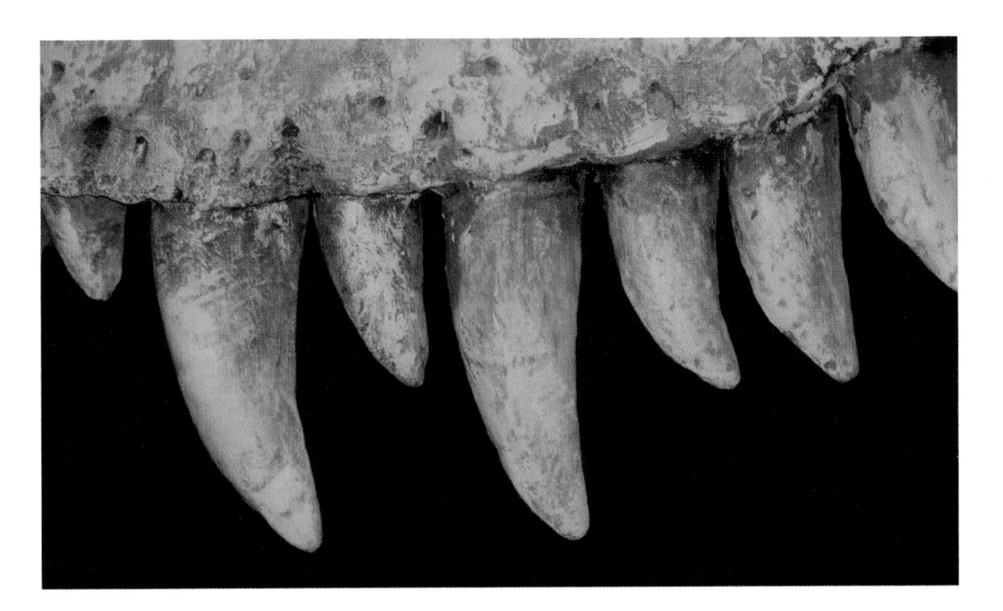

In a manner similar to today's sharks, many dinosaur species shed their teeth and grew new ones throughout their lives. Carnivorous dinosaurs had conical teeth with large gaps between them, and *Tyrannosaurus rex* had banana-shaped teeth with serrated edges (iStock).

As to how their teeth were aligned, there were many differences. In humans, most of the teeth are aligned so we can chew plants; the canines represent our carnivorous past. Dinosaurs had all different kinds of alignment. Some were similar to today's carnivorous crocodile reptiles; others had teeth perfect for gnashing tough plants. In other words, similar to modern animals, it all depended on what the dinosaur ate.

The plant-eating dinosaurs had much different teeth than carnivores, mainly because of the plants the herbivores had to eat. Meat can be readily cut up by sharp teeth, swallowed in large chunks, and digested easily in the gut. But living on a diet of plants is a much harder proposition. The cellulose found in plant tissue is much tougher than meat. Thus, plants must be cut into small pieces and thoroughly ground down by the teeth, a process that prepares the plant tissue for digestion in the gut. Once in the gut, microbes further break down the tough plant tissue in a long and rough process for the plant-eaters.

What can scientists tell from the **shape** of **carnivorous dinosaurs' teeth**?

The teeth of carnivorous dinosaurs were much different than those of herbivores. In general, a carnivore's teeth had large gaps between them; the teeth acted as daggers, powered by the force of the jaw muscles and the dinosaur's weight. The teeth in a carnivorous dinosaur were different sizes, since new teeth were continually growing to replace those lost or broken, mostly from biting into bone or even fighting with other carnivores.

A typical large carnivore, such as an *Allosaurus,* had backward curving, knifelike teeth. Each tooth had serrations on the front and back edges. Paleontologists

Did some dinosaurs have self-sharpening teeth?

Yes, some dinosaurs did have self-sharpening teeth, such as the ornithopods and ceratopsians. Both were members of the Cerapoda group, all of which had cheek teeth with thickened enamel on one side only. The other side of the teeth—the actual surface used to chew the plants—had only the softer dentine. As the animal chewed, the upper and lower teeth ground against each other, with the softer dentine wearing away faster than the harder enamel. This made these dinosaur teeth self-sharpening.

theorize that a large carnivore like the *Allosaurus* would practice what is called macropredation. It would run into its victim, mouth open as wide as it could to drive the teeth in as far as possible. Closing its jaws, it would begin to jerk its powerful neck, ripping off a huge chunk of meat. The animal would then swallow this chunk whole, letting its digestive system take care of the rest. It would continue tearing off pieces of the victim until it was full.

What was **unusual** about the **teeth** of a ***Tyrannosaurus rex***?

The teeth of a *Tyrannosaurus rex* were unusual in their shape. Most carnivorous dinosaurs had somewhat flat teeth, similar to razor-edged knife blades. However, a *Tyrannosaurus rex* had teeth shaped like giant spikes—almost like large, sharp bananas with serrations.

Are there any known **toothless theropod dinosaurs**?

Yes, there are several toothless theropod dinosaurs currently known, all from the group called the coelurosaurs. They include the ornithomimosaurs and oviraptorosaurs. Toothless dinosaurs had beaks that they used in place of teeth. In addition, some had strong jaw muscles that—when combined with short, deep skulls—could crush their prey.

Were the **arms** of some **carnivorous dinosaurs** used for anything?

The arms of some carnivorous dinosaurs were equipped with ferocious claws. The claws allowed the animals to grasp prey while the animal ripped off large chunks of meat, or to hold down prey while they bit and slashed, trying to bring down the victim. Others, such as a *Deinonychus,* would hold down its victim with its long arms, positioning itself to slash the prey with its large, hind claws.

What were the **dromaeosaurids** and what type of **claws** did they have?

The dromaeosaurids take their name from the first dinosaur fossil of their type found: the *Dromaeosaurus,* or "running reptile." This group includes species such

The claws of predator dinosaurs were all impressive, but none more so than those of the raptors, whose name means "hunting with the claw" (iStock).

as the *Dromaeosaurus, Velociraptor, Deinonychus,* and the more recently discovered *Utahraptor*. These dinosaurs were fast running meat-eaters, with long legs and light bone structures. They exhibited a blend of features from both the carnosaurs and the coelurosaurs, with some unique characteristics of their own.

Probably the most well-known feature of the dromaeosaurids was their large, sickle-shaped claws on the second toe of their foot. These huge curved claws were probably used to tear and rip apart their prey. Dinosaurs from this group are known colloquially as "raptors." Overall, they were all thought to be vicious pack hunters prey would not want to meet in the tall grass.

What is the **largest raptor claw** found to date in **North America**?

The largest raptor claw found to date in North America belongs to a *Utahraptor*. This dinosaur had a length of about 20 feet (6 meters) and a height of about 7 feet (2 meters); the estimated weight of this animal fell between 1,000 to 1,700 pounds (454 to 771 kilograms). It had large, crescent-shaped claws on its forefeet and hind feet, with the two large foot claws measuring 12 inches (30.5 centimeters) in length. While the dinosaur was alive, the claws were covered with a sheath of keratin and were somewhat larger (they probably shrunk from being fossilized).

What is the **largest raptor claw** found to date in the world?

To date, the largest raptor claw found in the world belongs to the *Therizinosaurus,* which had claws that measured around 28 inches (70 centimeters) long. This theropod beast lived in the area now called the Mongolian Desert 75 million years ago. It was a long distant relative of other ferocious meat-eaters like the *Velociraptor* and the *Tyrannosaurus*. But it's unknown how it used these huge claws because no one has ever found a complete skeleton of the creature. Scientists speculate that the claws may have been used for defense, mating displays, or—because it is thought that these theropods differed from most by eating primarily vegetation—even to reach treetops to forage on leaves.

DINOSAUR METABOLISM

Why is the **debate** over **dinosaur metabolism** so important?

The debate over dinosaur metabolism is important because it determines the way we visualize the behavior of these animals. If dinosaurs were cold-blooded (similar to today's reptiles) they were probably mostly sluggish, with only occasional bursts of quickness. Plus, they would not have been very smart. They probably would have spent most of their time basking in the sun, moving only to obtain more food, similar to the behavior of modern crocodiles. On the other hand, if dinosaurs were warm-blooded (similar to today's mammals) they were probably active, social animals. They would have been quick, alert, and intelligent. They would have spent much of their time actively grazing, similar to modern antelope, or hunting in packs, similar to lions and wolves.

Who first **proposed** that dinosaurs were **warm-blooded**?

In the late 1960s and early 1970s, paleontologists John H. Ostrom (1928–2005) and Robert T. Bakker (1945–) first suggested that dinosaurs were not sluggish, stupid, cold-blooded animals. Their work paved the way for the theory that many of these animals were actually agile, dynamic, and smart. In 1969, Ostrom published a description of the *Deinonychus,* a Cretaceous period carnivorous dinosaur. Based on his study of the creature, he theorized that dinosaurs may have been warm-blooded. In 1975, Bakker summarized his ideas about dinosaur endothermy in an article published in *Scientific American*. This set off a new era in dinosaur paleontology that continues today, especially in advancing ideas on how dinosaurs truly regulated their bodies' metabolism and heat.

Why did **Robert T. Bakker** believe **dinosaurs** were endothermic homeotherms **(warm-blooded)**?

Some of the reasons paleontologist Robert T. Bakker (1945–) gave for dinosaurs being endothermic homeotherms, or warm-blooded, are:

1. Dinosaurs had complex bone structures (with evidence of constant remodeling), a feature of modern mammals, not reptiles.
2. Dinosaurs had an upright structure similar to birds and modern mammals.

Do we know what type of metabolism dinosaurs had?

Unfortunately, we currently do not know too much about dinosaur metabolism. There is a wealth of conflicting evidence, most of it indirect. Some of the problem may be that various types of dinosaurs had different metabolisms. For example, the large plant-eating dinosaurs might have needed a much different metabolism than the smaller, quick theropods. Paleontologists are currently divided on their opinions of dinosaur metabolism, generally falling into three camps. Some interpret the available evidence as meaning dinosaurs were indeed cold-blooded. Others firmly believe that dinosaurs were warm-blooded. And the last group thinks dinosaurs had a unique physiology that included a combination of both cold- and warm-blooded traits.

3. Dinosaurs, at least the small theropods, had, based on the evidence to date, active lifestyles.
4. Predator-versus-prey ratios were closer to that of modern mammals than reptiles.
5. Dinosaurs have been found in polar regions, a climate that cold-blooded animals would not likely inhabit.

What can **dinosaur bones** tell us about their **internal heat regulation**?

Dinosaur bones, like modern reptile bones, often show signs of not growing—as if these animals went through periods of little or no growth. One of the reasons for a lack of growth could be hibernation during periods of seasonal cold, indicating that the animals used an ectothermic method of heat regulation. Mammals and birds, on the other hand, are endothermic, and show no lines of arrested growth. Thus, the presence or absence of these lines give paleontologists possible clues about the way dinosaurs regulated their internal temperature.

What kind of **heart** did **dinosaurs** have?

Fossil evidence of the soft internal parts of dinosaurs, including the heart, is sadly lacking. But based on indirect evidence—and one heart from a dinosaur called "Willo"—paleontologists have extrapolated that dinosaurs had a divided heart capable of keeping the blood at two different pressures.

Dinosaur bone tissue also shows evidence of blood vessels. Therefore, a heart was necessary to drive the circulatory system, sending blood around the body. Dinosaurs with extremely long necks (such as the large sauropods) and those with heads held upright needed a high blood pressure. This would allow the blood to pump all the way to the brain when the animal was reaching for food. But such a blood system would have too high a pressure to safely circulate to the lungs for oxygenation. Thus, scientists believe dinosaurs probably had a divided heart capable of supplying blood at two different pressures into two separate circulatory systems.

Hunting theropods like this *Tarbosaurus* required larger brains and more active metabolisms in order to catch their prey (Big Stock Photo).

What does **theropod dinosaur brain size** tell us about their **metabolism**?

Unlike most of the other dinosaurs, some of the smaller theropods (carnivores) had large brains relative to their body size. The brains were equivalent to those found in similar-sized mammals. To function properly, large brains need a steady temperature and constant supply of food and oxygen, all of which could indicate a *potential* for the higher metabolism associated with warm-bloodedness.

Do **dinosaurs' noses** indicate they were **not warm-blooded**?

Some scientists believe that dinosaurs' noses may tell us something about dinosaur metabolism. Respiratory turbinates are small scrolls of bone or cartilage in the nose covered with membranes; the absence of these bones in dinosaurs' noses is probably a good indication that the animals were not warm-blooded. These turbinates are found in all warm-blooded animals, having evolved independently in mammals and birds; no known cold-blooded animals have them.

Warm-blooded animals breathe quite rapidly. The warm, exhaled air passes over the turbinates and cools, causing the moisture in the air to condense out onto the membranes. In turn, this prevents dehydration.

If dinosaurs were truly warm-blooded, they would need to have respiratory turbinates to prevent dehydration. Scientists recently used computer-aided tomography (CAT) scans to study dinosaur fossil skulls for any signs of respiratory turbinates. So far, the remains of *Velociraptor* and *Nanotyrannus* do not show any signs of these structures. Further analysis is planned to study fossil skulls from all the major groups of dinosaurs.

DINOSAUR SIZE

What was the **purpose** of the **different sizes** and **shapes** of dinosaurs?

Similar to today's animals, the different sizes and shapes were the result of adaptations to the dinosaurs' surrounding environments. In particular, the dinosaurs were probably typical of most animals: they needed to adapt to the prevailing conditions and changing food supplies in order to survive. Many times, these adaptations took the form of certain sizes and shapes, and probably even colors.

Was there an **upper limit** to the **size** of a **dinosaur**?

The answer to this question probably depends on the availability and locality of the dinosaur's food supply. In general, to support a larger weight, the bone sizes also increase, or else they would literally break under the animal's own weight. Plus, as the bones become thicker to support the increasing weight, the animal would become more and more cumbersome, limiting its ability to obtain food. Thus, for each dinosaur species, there was probably a definite limit to its size.

What was the **average size** among the **dinosaurs**?

The popular conception of dinosaurs is one of hugeness. But dinosaurs came in all sizes, and shapes were extremely diverse, much like today's birds. They ranged in size from the gigantic sauropods, like *Brachiosaurus,* to small, chicken-sized ones like *Compsognathus,* and every size in between. Because we have found relatively few fossils—compared to how many dinosaurs scientists believed once inhabited Earth—it is hard to give an "average" size for dinosaurs.

How do scientists determine the **weight** of a **dinosaur**?

It's not easy to determine the weight of a dinosaur. Scientists can only estimate an animal's weight by looking at the bones of the animal.

One method of determining a dinosaur's weight is by studying the cross-sectional area of a limb bone. This way, scientists can estimate the weight borne by the limb. But it's not only the weight times four legs that results in an estimate; it's also the position of the legs, the posture of the animal, the tissue and flesh on the legs, and the limb shape. Some scientists have tried to extrapolate the weight of a dinosaur by comparing the animals to modern living species, but apparently there is really no linear relationship between the two.

Kids visiting Dinosaur Park in Münchehagen, Germany, stand beneath a replica of the Seismosaur (iStock).

Which **dinosaurs** were the **biggest**?

There are so many claims to which animals are the largest carnivorous and herbivorous dinosaurs that it's difficult to keep up. The favorite contender for a carnivore is the famous *Tyrannosaurus rex,* the Cretaceous period carnivore found in North America and Asia. It measured over 40 feet (12 meters) in length. Several other challengers are the *Giganotosaurus* of South America, and the *Carcharodontosaurus* of North Africa, two huge meat-eaters. There is also the *Spinosaurus,* a dinosaur measuring up to 52 to 59 feet (16 to 18 meters) long, which is often declared the largest known carnivorous dinosaur.

Some scientists also give the lead to a certain fossil of a *Tyrannosaurus* whose huge pubis bone was found in Fort Peck, Montana, in 1997. This creature was so massive that scientists have given the fossil its own name: *Tyrannosaurus imperator*. This tyrannosaur's pubis bone measured 52.4 inches (133 centimeters) long; the pubis of *Giganotosaurus* was only 46.5 inches (118 centimeters) in length, which would make the Montana *Tyrannosaurus* about 15 to 20 percent larger than any other known meat-eating dinosaur.

The winner of the largest herbivore is even more confusing. One of the best contenders includes a massive herbivorous dinosaur called *Argentinosaurus huinculensis,* a South American sauropod of the Titanosauridae family that measured between 130 and 140 feet (40 and 42 meters) long. Another contender is the second-largest sauropod so far found, the *Paralititan* (meaning "tidal Titan"), a titanosaurid sauropod found in Egypt that lived about 100 million years ago. Also, *Seismosaurus hallorum,* which scientists believe was related to the diplodocus, weighed upwards of 100 tons (90 metric tons) and was 120 feet (almost 40 meters) long. A complete skeleton of this huge beast was discovered in 1985 in New Mexico.

No doubt, the hunt for the largest carnivore and herbivore—and dinosaur—will continue. Paleontologists will find new dinosaur bones, and one of them may one day prove to be the largest dinosaur ever known.

What are the **smallest dinosaur fossils** found to date?

There is some disagreement as to the smallest dinosaur fossil yet found. To date, the best contender for the smallest is the *Microraptor* ("little plunderer"), a bird-like (crow-sized) dinosaur from China, a coelurosaurid theropod about 16 inches (40 centimeters) long. Another vote for the smallest adult dinosaur fossil found to date is the *Compsognathus* ("pretty jaw"). This animal was slightly larger than a turkey, with a total length of approximately 3 feet (1 meter), and weighed approximately 6.5 pounds (2.9 kilograms). This small carnivore, nicknamed "Compy," lived during the Jurassic period and was a fast-running, agile predator that probably subsisted on insects, frogs, and small lizards.

There was once another claim for smallest dinosaur fossil: the *Mussaurus,* or "mouse lizard," was found in 1979 in South America. Once thought to be the smallest dinosaur, it is now known that the *Mussaurus* fossils were actually hatchlings of *Coloradisaurus,* which, when fully grown, would be larger than a *Compsognathus*. Their eggs were only 1 inch (2.54 centimeters) long; the fossil hatchlings were only 7.8 to 16 inches (20 to 40 centimeters) long.

Where have **dwarf dinosaur fossils** been found?

Dwarf dinosaur fossils have been found in Hateg, Romania. During the Late Cretaceous period, much of the land area of eastern Europe was inundated by the waters of the Tethys Sea. Thus, the land existed in the form of islands.

Dinosaurs, along with other animals and plants, were isolated on these islands, cutting the flora and fauna off from other larger landmasses. Over time, the dinosaurs on these islands became smaller in response to the limited ecological environment. For example, the *Telmatosaurus,* a primitive hadrosaur found in Hateg, was about 15 feet (5 meters) long, and weighed approximately 1,10 pounds, or just over a half ton (450 kilograms). This is about one-third the length and one-tenth the weight of other *Telmatosaurus* fossils found in other parts of the world. The larger dinosaurs were able to take advantage of greater territories and habitats, growing much more than their smaller, island-bound cousins.

Which **dinosaur** had the **longest neck** of any animal known?

Although the true longest neck is highly debated, it is thought that the *Barosaurus,* or "heavy lizard," had the longest neck of any known dinosaur. The reason for the debate is that fossils of the *Barosaurus* are some of the rarest known. Because there are not many other specimens to back up the longest-neck claim, many scientists do not believe this animal is the winner. Even though there is still debate, the *Barosaurus,* which was related to the *Diplodocus,* did have an enormously long neck that was thought to be longer than the *Brachiosaurus*. Fossil remains of the *Barosaurus* have been found in the western United States and in Africa. The only mounted skeleton in the world of *Barosaurus* is found in New York City at the American Museum of Natural History; it is depicted rearing up on its hind legs, as if confronting a predator.

Most of the longest necks belonged to the sauropod (herbivore) dinosaurs, creatures that probably needed the longer necks to reach food in high tree branches. Other major contenders include the *Brachiosaurus,* a sauropod that reached a height of 40 feet (12.2 meters), with much of that height a combination of its long neck and front legs. Still another is the *Mamenchisaurus,* a sauropod with a 33-foot- (10-meter-) long neck.

To date, the longest neck in relation to its body belongs to the *Erketu ellisoni,* a sauropod with a neck more than 24 feet (8 meters) long. It lived in what is now Mongolia's Gobi Desert about 120 to 100 million years ago.

What was possibly the **longest predatory dinosaur**?

Scientists believe that a predatory dinosaur called *Spinosaurus aegypticus* may have been 52 to 59 feet (16 to 18 meters) long, making it one of the largest predators, if not the largest. The animal's long spine evolved during the Early Cretaceous period. (In fact, one vertebra of a *Spinosaurus* is taller than the average human.) The tall spines formed a skin-encased, sail-like structure; some scientists suggest that the long spine was needed to hold the sails. No one knows the function of the structure, but there are some theories. One states that, because the animals lived in the tropics close to sea level, the sails may have been used to cool off the animals. Another idea is that the sails were used for attracting potential mates, or for scaring off potential rivals or other dinosaur predators.

DINOSAUR BEHAVIOR

EATING HABITS

What did **dinosaurs eat**?

Based on the popular representations of dinosaurs in the media, these large, vicious creatures could eat anything they wanted—and some probably did. In reality, there is very little direct evidence of what specific dinosaurs truly ate. But, based on rare evidence and other factors, paleontologists have made some assumptions about dinosaur diets.

Since dinosaurs lived on our planet for around 150 million years, they must have slowly adapted to the changing flora and fauna. Overall, there were apparently two major, and one minor, types of dinosaurs: herbivores, carnivores, and omnivores, respectively. Most of the dinosaurs were herbivores, or animals that ate available plants; the carnivorous dinosaurs, of course, ate animals, including other dinosaurs. Very few dinosaurs were omnivores—animals that ate both meat and plants.

What **food** was available to the **dinosaurs**?

The food available to the dinosaurs gradually evolved over the millions of years of the Mesozoic era, just like the dinosaurs themselves. Most of the dinosaurs ate plants. Fossil evidence of pollen and spores indicates that there were hundreds to thousands of different types of plants growing during the Mesozoic era, most with edible leaves. Some examples of possible dinosaur delicacies included ferns, mosses, horsetail rushes, cycads, ginkos, and evergreen conifers like pine trees and redwoods (grass had not yet evolved). Toward the end of the Mesozoic era, with the advent of flowering plants, fruits also became available.

For dinosaurs of the carnivorous persuasion, there was also a large selection of food choices in the Mesozoic "cafeteria." These included the early mammals, eggs,

What evidence do paleontologists use to determine the diets of dinosaurs?

Because the actual, physical evidence of dinosaur diets is so rare (the soft parts of a dinosaur are rarely fossilized), paleontologists have turned to indirect evidence to form some idea of the feeding habits of dinosaurs. These include coprolites, trackways, fossil assemblages, and tooth marks on bones.

turtles, and lizards that shared the landscape with the dinosaurs. Of course, there were also other dinosaurs to be hunted or scavenged.

Did **dinosaurs** need **water**?

It is safe to assume that dinosaurs, like all living creatures, needed water to live. They probably obtained water much like modern reptiles, either directly from a water source; or from their food, such as the plants they ate (herbivores), the animals they consumed (carnivores), or both (omnivores).

What are **coprolites**?

Coprolites are the fossilized droppings, or feces, of dinosaurs. Because of the soft nature of this fecal material, dinosaur droppings would often disintegrate before they had a chance to fossilize. If they dropped in the "wrong" place, such as ocean shore where the waves would wash the material away, the chances of the dung becoming fossilized were almost nonexistent.

The shapes and sizes of most coprolites are not readily distinguishable between animals. Thus, there are, at present, few coprolites unequivocally traced back to dinosaurs, but the ones that have been identified offer tantalizing clues to dinosaur diets. In particular, such coprolites give us an insight into what the animal was eating, how it ate, and what happened later in terms of digestion.

The preservation and subsequent fossilization of coprolites depended on a number of factors, including the organic content and amount of water present in the deposited feces. It also included, of course, where the animal dropped the feces and the method of burial—all keys to the formation of coprolites.

The feces of carnivorous dinosaurs were more likely to become fossilized than those of the herbivores because of their higher mineral content. These minerals were from bits of bone within the feces; in other words, from the consumption of other animals.

Are all **coprolites large,** especially those from the larger dinosaurs?

Not all coprolites are large. Individual coprolites can be small—less than 3 inches (10 centimeters) long—even though they came from a large dinosaur. A modern example of this phenomenon is the mule deer and elk of North America. These animals deposit many pellets that are less than a third of an inch (one centimeter) in size, even though these are relatively large animals.

In terms of dinosaurs, evidence for this phenomenon comes from a probable sauropod coprolite found in the Morrison formation of eastern Utah. Although this coprolite is about 16 inches (40 centimeters) in diameter, it is probably from a mass of smaller, individual pellets that merged together, probably because of a high water content.

Coprolites are actually fossilized dinosaur feces. Not a very glamorous subject for research, but paleontoligists actually discover a lot about what dinosaurs ate and more from these stones.This sample was from a young *T rex.* Discovered in Saskatchewan, Canada, it is about 17 inches (43 centimeters) long (U.S. Geological Survey).

What have paleontologists **learned** from **dinosaur coprolites**?

In coprolites thought to be from herbivorous dinosaurs, paleontologists have found quantities of cycad leaf cuticles, conifer stems, or conifer wood tissues, giving clues as to what these dinosaurs ate. Also, in some cases, the nature of the fragments shows these dinosaurs were well-equipped to chew up and digest the tough, woody food available at the time.

Several coprolites from carnivorous dinosaurs have also been found. One recent find was astounding. In 1998, near the town of Eastend, Saskatchewan, Canada, scientists found a huge coprolite that was almost as large as a loaf of bread. The 65-million-year-old coprolite measures 17 inches (43 centimeters) long and 6 inches (15 centimeters) thick and is one of the largest coprolites ever found. It is thought to be from a young *Tyrannosaurus rex*. The analysis of the coprolite also produced an intriguing suggestion: the remains look as if the *Tyrannosaurus* did not swallow the bones of its prey whole, but actually chewed and pulverized the animal's bones. This is contrary to what paleontologists have believed for a long time, which was that these carnivores "gulped and swallowed" their prey. But the verdict is still out on the subject until they find more such coprolites.

Did **dinosaurs urinate**?

No one knows if dinosaurs urinated, as the soft internal parts to indicate such an activity do not readily fossilize. But it is probably safe to assume they did, if modern reptiles and birds are similar to their ancient cousins. In fact, they may have excreted a solid form of urea, or guano, similar to modern birds and reptiles.

What do dinosaur **trackways** tell paleontologists about **dinosaur diets**?

Trackways, the multiple fossilized footprints of dinosaurs, have given paleontologists some idea of the feeding habits of many dinosaurs. For example, an Early Cretaceous period site in Texas, as well as a Late Cretaceous period one in Bolivia, show footprints of what appears to be a pack of theropods actively stalking a herd of sauropods. In a Cretaceous period site in Australia, fossilized tracks suggest a herd of over 100 small coelurosaurs and ornithopods stampeded as a large, single theropod stalked the group. In a Utah coal mine, Cretaceous period footprints of herbivorous dinosaurs cluster around fossil tree trunks, giving some indication of their foraging behavior.

What are **fossil assemblages** and what do they tell us about **dinosaur diets**?

Fossil assemblages are groups of fossils from different dinosaurs. For example, one assemblage found in the sandstones of the Gobi Desert is of a carnivorous *Velociraptor* intertwined with a herbivorous *Protoceratops*. The *Velociraptor*'s clawed feet were attached to its prey's throat and belly, while the *Protoceratops*' jaws had trapped the arm of the predator. This assemblage suggests the struggling dinosaurs, predator and prey, died together as a massive sandstorm overcame them.

In fifteen sites in Montana, major fossil assemblages have been found: the teeth of the carnivorous dinosaur *Deinonychus* found in association with the fossil remains of herbivorous dinosaur *Tenontosaurus*. Dinosaur teeth were continually shed, as new ones grew in, and vigorous biting could have increased tooth loss. In fact, there is lack of *Deinonychus* teeth found with the remains of other dinosaurs, leading paleontologists to conclude that the *Tenontosaurus* was the favorite prey of this predator.

What do **tooth marks** tell us about **dinosaur diets**?

Another piece of evidence used by paleontologists to determine dinosaur diets are tooth marks. Most of the grooves or punctures found associated with fossilized dinosaur bones were the result of attacks by carnivorous dinosaurs. Most of the time, however, this evidence does not reveal whether the victim was actively hunted or scavenged—and except in certain cases, the identity of the predator cannot be determined.

In one instance, the spacing of the scoring found on the bones of an *Apatosaurus,* a herbivore, matched the spacing of teeth from the jaw of an *Allosaurus,* a carnivore. In another case, dental putty was used to make molds of puncture marks found in a *Triceratops*'s (an herbivore) pelvis, and an *Edmontosaurus*'s (another herbivore) phalanx. The resulting molds nicely matched the fossilized teeth of the perpetrator, a carnivorous *Tyrannosaurus*. In one rare instance, a *Tyrannosaurus* tooth was found stuck in the fibula of an herbivorous hadrosaur, *Hypacrosaurus*. Of course, in the most obvious cases (and the most rare), predators could be identified if the fossil bones—and teeth—of the two animals were locked in mortal combat.

To what **group** did all the **meat-eating dinosaurs** belong?

All of the carnivorous, or meat-eating, dinosaurs belonged to the theropods, or bipedal carnivores. These dinosaurs, along with the large, herbivorous sauropods, made up the saurischian dinosaurs. This group represents a wide range of dinosaur species—from the large *Tyrannosaurus* to the small *Compsognathus.*

Duck-billed dinosaurs such as this *Corythosaur* had special teeth within their unusual mouths for grinding up plants (Big Stock Photo).

What adaptations enabled **carnivorous dinosaurs** to **eat meat**?

All of these dinosaurs, known as the theropods, shared many adaptations specific to the catching, killing, eating, and digesting of meat. These animals had larger, sharper, and more pointed teeth than their herbivorous cousins; they were used to kill the victim and tear the flesh off the body. To power these teeth—and to break down the nutritious bone marrow from their prey—they needed strong jaws and muscles.

The carnivores also had clawed feet for slashing their victims, with the dromaeosaurids possessing the epitome of this adaptation: large, sickle-shaped foot claws. The theropods, being bipedal, had their arms and hands free to grasp their prey; their fingers often had claws used to slash and hold the victim. Being bipedal, they had the relative speed and agility to catch their prey, especially sick and ailing animals. It is thought by some scientists that the theropods had good eyesight, a keen sense of smell, and a large brain (in proportion to its body) to calculate hunting strategies.

What **adaptations** did the **herbivorous dinosaurs** have that enabled them to **eat plants**?

Some herbivorous dinosaurs did not chew at all, but merely swallowed whole the vegetation they pulled off a tree or bush. They had larger (and probably more rugged) digestive tracts than carnivorous dinosaurs in order to digest the tough, fibrous plants they ate. Some herbivores, such as the *Ankylosaurus,* even had fermentation chambers along their digestive tract in which tough fibers would be broken down by bacteria. In addition, some herbivores had gastroliths, or "gizzard stones," in their digestive tract, which would grind up the fibrous plants, helping to digest the material. (It is interesting to note that this method is similar to how birds swallow stones to grind up ingested matter in their digestive tracts.) These stones were deliberately swallowed, and are often found with fossils of herbivores. Both of these actions prepared the vegetation for digestion.

Other herbivorous dinosaurs, like the duck-billed hadrosaurs, had special teeth that would grind up the food before swallowing. Ceratopsians like the *Triceratops* had sharp teeth and powerful jaws that enabled them cut through tough plants. Still other herbivores had cheek pouches, apparently used to store food for later ingestion. They probably concentrated their meals on certain plants, especially the ancestors of the conifers, flowering plants, horsetails, ferns, and cycads that grow today.

What was an **omnivorous dinosaur**?

An omnivorous dinosaur was one that ate both plants and meat. There are only a few known omnivores among the dinosaurs, including *Ornithomimus* and perhaps *Oviraptor,* although new fossil finds may change scientists' opinion about the latter. A diet of an omnivorous dinosaur could include different types of plants, insects, eggs, and small animals. These omnivorous dinosaurs were probably rare, only eating this way out of necessity, such as when there was a sudden lack of meat or plants in their surrounding habitat. Others believe these dinosaurs were omnivores by accident, eating insects and small animals as they ate the plants around them.

Have any **fossilized stomach contents** of dinosaurs been found?

Although rare, some dinosaur stomach remains have been found over the years. The best examples are from carnivorous dinosaurs: The fossilized remains of a lizard (*Bavaisaurus*) were found in the gut region of a carnivorous dinosaur called *Compsognathus*—no doubt the dinosaur's last meal. In addition, *Coelophysis* fossils have been found with the fossilized remains of other *Coelophysis* dinosaurs inside the gut region, indicating these dinosaurs probably engaged in cannibalism. Whether this was active predation or scavenging has not been determined.

The stomach contents of herbivorous dinosaurs have not been as definitive, however, due to the organic nature of the material. One case was reported in the early 1900s, where the fossilized remains of an *Edmontosaurus* had been found with conifer seeds, twigs, and needles in the body cavity. However, it could not be determined if these were actual stomach contents, or just debris that had subsequently washed into the carcass.

Did other **animals prey** on **dinosaurs**?

Yes, and in 2005, a 130-million-year-old mammal fossil found in China was important evidence of this. A fossil of a cat-sized mammal known as *Repenomamus robustus* was found with a small dinosaur preserved in its stomach area. This was the first direct evidence that mammals preyed on dinosaurs.

DINOSAURS IN MOTION

What are some major **questions** associated with the **movement of dinosaurs**?

When paleontologists study the movement of dinosaurs, the most prevalent questions concern the postures that dinosaurs typically adopted, their speed of movement, both normal and maximum, and any hunting or herding behavior.

What **evidence** do we have concerning the **movement of dinosaurs**?

There are three main sources of evidence that help us to understand the movement of dinosaurs: homologies, analogies, and footprints.

Homologies are comparisons in the anatomical structures in organisms derived from the same such structures in a common ancestor. Because the vast majority of evidence we have about dinosaurs are their bones, the reconstruction of the skeleton and muscles can be very useful in helping to understand their movement. To make these reconstructions as accurate as possible, scientists use modern homologues. Unfortunately, the closest relations to the dinosaurs, such as the birds and crocodiles, have all evolved highly modified structures, making a direct comparison to the dinosaurs very unreliable.

In addition to homologies, scientists can observe the movement of modern animals with similar structures and probable behaviors (analogies). For example, the probable motion of ornithomimids is based on that of the ostrich because this modern bird has a similar structure; the movement of the large sauropods is modeled on that of modern elephants. But these analogies are only as good as the similar structure the scientist is examining. Unfortunately again, it may not be truly representative of the actual motion or behavior of the dinosaur. In other words, ostriches are not theropod dinosaurs, and elephants are not sauropods.

The third and most direct evidence for dinosaur movement is the trace fossils of footprints. These records give us a wealth of clues about the speed, gait, posture, and sometimes behavior of these long-dead animals. From the inferred motion represented by these footprints, paleontologists can sometimes obtain evidence of the animals' behavior. For example, the lack of tail drag marks shows a dinosaur with an erect posture; some trackways show that certain dinosaurs exhibited a herding behavior.

What is an **important factor** in determining the probable **motion of dinosaurs**?

One of the most important factors in determining the probable motion of dinosaurs is their posture. Without an accurate knowledge of the animals' postures, their

By analyzing anatomy and footprint fossils, and by drawing comparisons with today's animals, scientists have made various hypotheses on how dinosaurs like this *Deinonychus antirrhopus* may have moved (Big Stock Photo).

motion—whether obtained through the use of homologies, analogies, trackways, or a combination of these—can be misleading or downright wrong.

From the evidence of footprints, dinosaurs had an erect (or upright) posture, with their limbs directly under their bodies. This structure resulted in a motion similar to modern mammals, with the limbs held out to the sides and the upper bones nearly parallel to the ground. Surprisingly, this posture was more similar to modern mammals than lizards, which have a sprawling posture.

But there were also variations within this erect posture, leading to some very unique movements. Some dinosaurs were bipedal, walking only on their hind limbs; others were quadrupeds, walking on all four limbs. Other dinosaurs spent most of their time on all fours, but were capable of standing on their hind limbs. Still others reversed this action, spending most of their time on two hind limbs, but capable of moving on all fours. In the end, each had a distinct motion and behavior based on its respective posture.

What are **dinosaur trackways**?

Dinosaur trackways are fossil footprints of dinosaurs, which are found all over the world. These trackways developed as dinosaurs (and other animals) walked in the soft sediment or sand along the shorelines of beaches, rivers, ponds, and lakes. Almost immediately after the animals walked by, the tracks were quickly buried in sediment, eventually becoming fossil footprints (also called ichnotaxia). Because

What similar trackways were recently found in Scotland and Wyoming?

Similar trackways made by a three-toed dinosaur were recently found in Scotland and Wyoming—a rarity since scientists have never found almost identical dinosaur tracks on two different continents. The 170-million-year-old tracks from the Jurassic period were painstakingly measured, revealing definite similarities between both fossil tracks. Thus, researchers believe the tracks were made by the same type of dinosaur species. This may be true, since the areas of today's Wyoming and Scotland were much closer during the Jurassic—only a few thousand miles apart—making it possible for the species to migrate and live in what are now two very different places. But not all scientists agree, some saying that it is more likely that similar—but distinct—dinosaurs species evolved in these two areas because the creatures were at similar latitudes.

such places were good sources of water and food, including lush plants for the herbivores, and plenty of animals for the carnivores, they became natural pathways for all types of dinosaurs. (In fact, the branch of science concerned with the study of footprints is called ichnology.)

So far, the problem with dinosaur trackways is that it is impossible to tell what dinosaur made the footprints. Scientists can only give general an idea of the animals, such as if the tracks belonged to a biped or quadruped dinosaur, and often a sauropod versus a theropod. But that is usually all they can tell from a footprint.

Where was one of the **first dinosaur footprints** found in the **United States**?

The first fossilized dinosaur footprint found in the United States is attributed to Pliny Moody in 1800 in Massachusetts. The 1-foot (0.3-meter) print was uncovered at Pliny's farm—and it was initially thought to be from "Noah's Raven," as per the "Noah and the ark" story in the Bible. Other dinosaur footprints were also found in various New England quarries in the early 1800s, but they were thought to be unimportant. The majority of the tracks were consequently blown to bits in the quarrying process.

What **conditions** might lead to the best **footprints**?

Recent experiments show that the best footprints aren't made in fresh, new mud. Instead, the presence of a coating on top of older mud leads to the best tracks.

Fresh mud tends to be sticky and flows readily. Footprints made in this type of mud are poorly preserved and leave few little details. Fresh, sticky mud adheres to the feet, leaving at best a partial impression in the ground; as the foot is lifted, some of the surrounding mud flows into the footprint. Footprints made in this type of

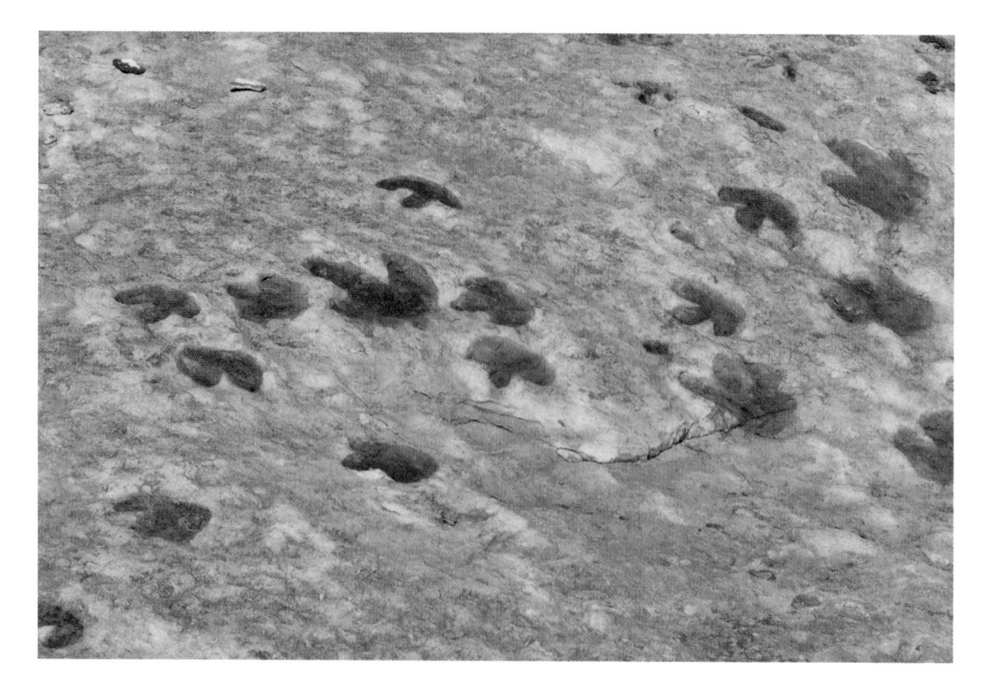

Under the right conditions, even the footprints of dinosaurs may be preserved for eons. Such finds can tell paleontologists a great deal about the anatomy and behavior of dinosaurs (iStock).

mud seem to be typical of most of the known dinosaur tracks, which is why they are so hard to associate with particular species.

On the other hand, older coated mud seems to be a much better medium for making and preserving finely detailed footprints. This type of mud is commonly found around ponds and is sometimes covered with a greenish coating of algae and bacteria. Experiments have shown that the coating acts as a binding agent. It keeps the muddy surface together and prevents flow into the footprint after the foot is lifted. It also acts as a parting agent, preventing the mud from sticking to the feet and resulting in deep prints with fine details. A third benefit to the coating is that it slows down the drying of the mud, allowing the prints to be formed.

Tracks made in this coated mud can be deep, very clear, and well-preserved, with plenty of anatomical detail. The most well-known dinosaur footprints were probably made under these conditions.

What are some of the **largest** dinosaur **trackways** in **North America**?

Some of the largest dinosaur trackways are called megatrack sites, where footprint-bearing rock can extend for hundreds or even thousands of miles. Several Jurassic and Cretaceous periods sites in North America have such trackways. For example, tracks in the Entrada sandstone beds in eastern Utah (Middle Jurassic) cover about 116 square miles (300 square kilometers); the density of the prints is estimated to be between 1 and 10 per 10.8 square feet (1 square meter). Another is a 150-mil-

Where and what is the Dinosaur Freeway?

The Dinosaur Freeway is a large trackway of dinosaur footprints extending along the Front Range of the Rocky Mountains, from around Boulder, Colorado, to eastern New Mexico. In the Middle Cretaceous period, this area was a coastal plain with a wide shoreline, a good source of water and food.

lion-year-old dinosaur trackway in the Purgatoire River Valley, Colorado. The area was once home to a large freshwater lake, allowing dinosaurs to walk through the mud along the edge, and leaving behind vast trails of footprints. Today, over 1,300 of these now-solidified-in-rock footprints are exposed at the Picketwire Canyonlands dinosaur trackways.

What are some of the **largest** dinosaur **trackways outside North America**?

Other dinosaur trackways have been discovered around the world. For example, at the Lagosteiros Bay in Cabo Espichel, Portugal, there is a large trackway complete with multiple tracks from the upper Jurassic. The impressions represent numerous dinosaurs, including a sauropod that limped and examples of sauropods that moved as a herd.

What do **dinosaur tracks** tell us about **locomotion** of some dinosaurs?

Dinosaur trackways can tell us a few things about dinosaur behavior. For sauropods, the tracks are usually of more than one creature and head in the same direction, indicating a social herding behavior or even a migration. Some trackways include footprints of large theropods; some prints indicate a pack behavior for stalking large sauropods.

Dinosaur trackways confirm that certain dinosaurs walked and ran on all four legs (quadrupeds) and others on two legs (bipeds). The tracks also show that some dinosaurs walked in an erect fashion, putting one foot almost directly in front of the other. In addition, some dinosaurs quickly ran, or walked slowly, probably depending on whether the animals were browsing, wading, trotting, running after prey, or running from predators. Another interesting observation: so far, very few tail marks—indicating the dinosaurs dragged their tails along behind them—have been found in trackways. Because of this, scientists believe most dinosaurs probably held their tails erect.

Where did a **dinosaur stampede** take place?

Evidence of a stampede was discovered in 1960 in Australia. There, in the Lark Quarry Environment Park, south of Wilton, on the eroded edge of the Tully Range, are hundreds of dinosaur footprints preserved in rock. The footprints were made as dinosaurs walked in mud around a prehistoric lake.

Typical for animal life, most of the tracks were made as large carnivores hunted for prey along the edge of the lake. In particular, large carnosaurs trapped groups of coelurosaurs and ornithopods; in one instance, a carnosaur attacked an unfortunate animal, pursuing its victim along the muddy shore—and causing the rest of the surrounding dinosaurs to stampede in panic. And although there is no longer a large lake with carnivorous dinosaurs, the area still holds a bit of danger. Because of the rough terrain, it is not easy to reach this park.

How do paleontologists determine **speeds of dinosaurs** using the **trackways**?

Although it is difficult to tell the type of dinosaur that made a track, scientists can tell the relative speed of the animals as they moved along the trackways. By measuring the distance between the footprints and the size of the tracks, they can tell that some dinosaurs ran much faster than first assumed. In other words, these tracks show that the old idea that dinosaurs were slow and sluggish wasn't always true.

What are some **calculated speeds** of dinosaurs?

Using measured stride and footprint lengths, scientists have calculated the speeds of over 60 dinosaur species. One of the real life factors to remember while doing these calculations is the gait of the dinosaurs when they made the tracks. The difference between walking and running and the transitions between these movements had to be kept in mind. Another factor was the actual leg bone lengths of the dinosaurs, which helped determine the reasonable speed types for the various dinosaurs. For example, the similar lengths of the femur and tibia bones in the legs of a *Tyrannosaurus rex* suggest a slower attainable speed than that of the ornithomimids, which had a shorter femur and longer tibia. With all these caveats in mind, some examples of the calculated speeds of dinosaurs include:

Dinosaur Speeds

Dinosaur	Maximum Speed (mph/kph)
Ornithomimids	37/60
Small theropods and ornithopods	25/40
Ceratopsians	16/25
Large theropods, ornithopods	12/20
Sauropods	7.5–11/12–17
Armored dinosaurs (e.g., ankylosaurs, stegosaurs)	4–5/6–8

Humans can run at speeds of about 14 miles (23 kilometers) per hour.

How are **dinosaur speeds calculated**?

Paleontologists first measure the distance between footprints in a trackway (stride length), as well as the length of the footprints themselves. The estimated leg length is then determined by multiplying the footprint length by a known constant. For

What does the Red Gulch Dinosaur Tracksite reveal?

The study of the Red Gulch Dinosaur Tracksite in Wyoming—located on the United States Department of the Interior's Bureau of Land Management (BLM) land—has revealed a great deal about the dinosaurs that made the tracks; more about the environment during the middle Jurassic period in this part of the world; and how the sedimentary layer was deposited. All of these are clues that will help to reconstruct the world of these animals, from large to small dinosaurs. Scientists continue to determine the specific dinosaurs that made the tracks, as well as whether or not the animals were bipedal or quadrupedal. In addition, the footprints have shown evidence of behavior patterns, including whether these dinosaurs were solitary or lived in family groups.

It is possible to visit the site, but much of the surface is administered by the BLM. So while hobbyists may collect petrified wood, invertebrates, and plant fossils, vertebrates uncovered at the site are kept in the public trust through the BLM's collecting permit process. (For more about the Tracksite, visit their Web site at http://www.blm.gov/wy/st/en/field_offices/Worland/Tracksite.html.)

example, the constant for theropods is approximately 4.5. The relative stride length is determined by dividing the measured stride length by this estimated leg length. The resulting number would be referenced to a standard graph to obtain a dimensionless speed for the particular dinosaur. To get a real world speed, this dimensionless speed is used, along with the estimated leg length and the acceleration of gravity. This results in a more familiar number: speed in miles per hour or kilometers per hour.

What was the **fastest dinosaur** based on the **trackways**?

It is difficult to name the fastest dinosaur based on the few trackways found, but some information has been gathered by analyzing the tracks. The speediest dinosaurs were probably the small, bipedal carnivores, especially those with long, slim hind limbs and light bodies. These swift dinosaurs probably didn't run any faster than the fastest modern land animals. One carnivorous dinosaur called an *Ornithomimus* is thought to have run about 43 miles (70 kilometers) per hour, which is about the speed of a modern African ostrich.

Did **dinosaurs** travel in **herds**?

Yes, some dinosaurs did apparently live and travel in herds, probably because there was "safety in numbers." Scientists have deduced this behavior based on dinosaur trackways and huge collections of dinosaur bones that indicate massive kills (places in which large amounts of dinosaurs bones are found in one place).

In particular, many herbivores apparently traveled in herds, based on the multiple tracks left along the dinosaur trackways. The tracks also show that many herbivores held the young in the center of the herd (similar to elephant and bison herds), most likely to protect them.

Some dinosaur fossils have been found in massive collections, indicating many dozens of animals were killed in one spot. Some scientists believe such collections of animal bones show the creatures exhibited a herding behavior. In many cases, while in the herd, these animals were swiftly killed off, perhaps from a major flood, volcanic action, or a huge sandstorm. For example, the bone beds of about 100 *Styracosaurus* dinosaurs, a herbivore, have been discovered, as have dinosaur bones that represent dozens of *Protoceratops* and *Triceratops* in herds.

One particular herbivore called a *Maiasaura* (a hadrosaur) is also thought to have lived in herds and probably returned to the same nesting grounds every year. Fossil bones of these animals were found in a huge group of about 10,000 animals in Montana. The animals all died suddenly, apparently when a volcano erupted, smothering the animals with volcanic gases and covering the creatures with a thick layer of ash.

Did **dinosaurs migrate**?

Yes, certain dinosaurs apparently migrated in a manner similar to certain animals today. They probably migrated for the same reasons, too, such as to seek new food sources as the seasons changed, or to reach mating grounds. Scientists have deduced patterns in migrating behavior based on trackways, as well as huge collections of bones that indicate massive kills.

Did **dinosaurs hunt** in **packs**?

Yes, paleontologists speculate that certain carnivorous dinosaurs exhibited a social behavior called pack hunting. The large theropods, like *Tyrannosaurus* and *Giganotosaurus,* show some evidence of hunting in packs similar to modern-day lions.

In a recent discovery in Argentina, scientists found a huge collection of dinosaur bones of *Giganotosaurus*—a carnivorous dinosaur that grew to 45 feet (13.7 meter) long and weighed about 8 tons (7 metric tons). The bones of four or five *Giganotosaurus* dinosaurs indicate that they died together on the Patagonian plains, swept away by a fast-flowing river. The bones show that two of the animals were very big, but the others were smaller. Scientists believe that this shows that there was some kind of social behavior, such as pack hunting. Each animal within the group would have different characteristics, giving the pack a great range of capabilities, such as going after smaller and larger animals.

Other evidence shows that at least some of the dromaeosaurids, or "raptors," engaged in pack hunting. When the first fossils of a dinosaur called *Deinonychus* were found—a 6-foot- (1.8-meter-) tall, 9-foot (2.8-meter-) long Cretaceous period predator of western North America—the remains of many of these carnivores were clustered near the body of a large herbivore, a *Tenontosaurus*. Paleontologists theorized that these predators perished during the struggle with the larger dinosaur, indicating that the hunt was being conducted by a group.

The *Apatosaurus*'s long, strong tail was used for balance, many scientists believe, and might have even been used to help it rear up on its hind legs so as to better reach food or, perhaps, appear even more intimidating to predators (iStock).

Did any **dinosaurs** climb or live in **trees**?

No known true dinosaurs climbed or lived in trees. At one time, the foot bones of the *Hypsilophodon,* a small herbivorous ornithopod, were thought to have the big toe facing opposite to the other toes, similar to a bird's foot. Scientists speculated, based on this erroneous assumption, that this dinosaur lived in trees, perching on branches in a way similar to today's Australian tree kangaroo. When the true foot bone structure was discovered, however, it was realized that *Hypsilophodon* was a swift land runner, using its speed and agility to escape predators and not to live in trees.

There are some scientists who speculate that some of the feathered dinosaurs, such as the *Microraptor* from China, may have glided from tree to tree or used its extended toe to grasp tree branches. The animal may have adapted this way to escape predators or even to find a better meal. But more evidence needs to be found before researchers can say dinosaurs lived or climbed in trees.

Did any **dinosaurs fly**?

No known non-avian (non-bird), flying dinosaur has ever been discovered, but there were creatures that flew at the time of the dinosaurs. The first and most prolific were the pterosaurs. These contemporaries of the dinosaur lived throughout the

Mesozoic era and appeared to be close relatives of the dinosaurs, evolving from the archosaurs. The dinosaurs dominated on land; the pterosaurs dominated the air.

The connection of dinosaurs to birds is still highly debated. But if dinosaurs and birds are truly so closely related, then it could be said that avian dinosaurs, or birds, would eventually evolve, take to the air, and endure today.

What was the **purpose** of the giant sauropod ***Apatosaurus's*** (formerly known as Brontosaurus) **tail-tip**?

Some scientists believe the *Apatosaurus* may have moved its 45- foot (14-meter-) long tail like a bullwhip, with its 6-foot- (1.8-meter-) long, skinny tip creating a loud crack. Recent computer models have compared the movement of a model *Apatosaurus* tail with the motion of a bullwhip. These studies show it is not only feasible, but the relatively slow motion at the base of the tail would translate into supersonic speeds at the tip. This would have created a loud sonic boom of about 200 decibels—much louder than the 140 decibels of a jet taking off.

In the past, paleontologists thought the tail of sauropods like *Apatosaurus* were mainly used for balance, or to swat rivals. But the tip of the tail contained tiny, fragile bones that could have easily broken in a fight. Now, some scientists think the loud crack from the tail tip could have been used to scare away predators, establish dominance in a herd, resolve disputes, and even to attract a mate.

DINOSAUR BABIES

What do **dinosaur eggs look like**?

Scientists have collected fossil dinosaur eggs, sometimes finding more than a dozen in a nest-like area. The fossilized eggs are usually the color of the rock in which they are found. Similar to fossil dinosaur bones, their structures have been fossilized and replaced by minerals over time.

Although the eggs are fossilized, scientists have discovered that dinosaur eggs probably looked similar to those of modern birds, reptiles, and some primitive mammals. Most of the eggs were rounded or elongated, with hard shells. They contained an amnion, a membrane that kept the egg moist, a kind of "private pond" for the young animal growing in the egg. The eggs appeared to be similar in other ways, too. The surface of the shell allowed for the exchange of gases necessary for the young to survive (many of the fossilized eggs exhibit a mottled surface that indicates the shell pores), and the young would crack their way out of an egg when they were ready to enter the world.

No one really knows whether the majority of eggs laid by the dinosaurs were soft-, flexible-, or hard-shelled. The eggshell would have to have been relatively strong to support the weight of a brooding parent, or the overburden of nesting material, while still allowing for the exchange of necessary gases. Hard-shelled eggs

What is the largest dinosaur egg known?

The largest dinosaur egg to date is about 12 inches (30 centimeters) long and 10 inches (25 centimeters) wide; it may have weighed about 15.5 pounds. It is thought to be from a giant, 100-million-year-old herbivore called a *Hypselosaurus*. To compare, the prize for the largest bird egg (and largest flightless bird) on Earth belongs to the African ostrich, with eggs up to 6.8 inches (17 centimeters) long by 5.4 inches (14 centimeters) wide, and weighing up to 3.3 pounds.

had the best chance of fossilizing, so most shells that we find today may not truly represent all the eggs that dinosaurs laid, but just the hardest ones that survived.

Did **all** dinosaurs lay **eggs**?

As far as paleontologists can determine, all dinosaurs reproduced by laying eggs. The first fossilized dinosaur eggs were found in France in 1869, but not everyone agreed the eggs were from dinosaurs. Although it may seem somewhat obvious to us now, it took time before scientists agreed that dinosaurs nested and laid eggs. The proof was found in the 1920s in the Gobi Desert, where nests and eggs of a group of *Protoceratops* were found. Since that time, over 200 sites with fossil eggs of various dinosaurs have been found all over the world, including in the United States, France, Mongolia, China, Argentina, and India.

It should also be noted, though, that there are some modern reptiles that do not lay eggs outside their bodies; rather, their eggs remain inside them, where they hatch and then emerge as live young (this is called being ovovivaporous). Some scientists suggest that this is a result of adapting to colder climates. And although the idea is highly debated (and no real physical evidence has been found), some scientists speculate that polar dinosaurs may have reproduced in this way.

Where were the **first clutch** of dinosaur eggs **discovered**?

The first known clutch of dinosaur eggs were found by American naturalist Roy Chapman Andrews (1884–1960) in 1922 in the Gobi Desert, south of the Altai Mountains. The eggs were found in one of the most prolific fossil beds in the area, a sedimentary layer known as the Nemget formation, which is also known for the more than 100 fossil skeletons of the small horned dinosaur *Protoceratops*.

Have **other clutches** of dinosaur eggs been **discovered**?

Yes, there are hundreds of sites around the world that have large clutches of dinosaur eggs. For example, in a remote area of Argentina, in northwest Patagonia (Auca Mahuida), the concentration of eggs was so rich that, in an area of roughly 300 feet by 600 feet (91 meters by 183 meters), scientists counted about 195 clusters of eggs.

All dinosaurs laid eggs, making a variety of kinds of nests. Paleontologists have discovered many eggs and nesting sites in recent years, some including the fossils of not only eggshells but also hatchlings and even the mothers' remains (Big Stock Photo).

What is one of the **smallest dinosaur eggs** known?

So far, scientists claim the smallest dinosaur eggs ever found are from China. The four eggs, two of which contain remains of embryos, are from a dinosaur that may also turn out to be the smallest known (the *Microraptor*). The egg is about the size of a goldfinch egg, measuring about 0.7 inch (18 millimeters) long, or roughly the width of a thumbnail.

Have any **dinosaur eggs** been found in a **dinosaur's belly area**?

Yes, dinosaur eggs have been found in the belly area of a fossilized dinosaur mother that is thought to be a two-legged oviraptor from China that lived between 100 to 65 million years ago. In 2005, scientists announced they had found a fossil of the first dinosaur eggs—complete with shells—in the body of the mother. Some researchers believe this shows how some dinosaurs laid their eggs in a clutch, and in a series of sittings (similar to modern birds), rather than all at once the way crocodiles and other living reptiles do.

What did a **dinosaur nest** look like?

Not all dinosaur nests looked alike. Many were simple pits dug into the soil or sand; others were more complicated, including deep, mud-rimmed nests lined with vegetation. Some dinosaurs even had a certain way of laying their eggs. For example, the *Maiasaura,* a herbivore, would arrange its eggs in a spiral, making sure to allow enough space between hatchlings to aid them in escape from predators. *Protoceratops* also apparently laid eggs in a spiral fashion.

What else has been **discovered** about **dinosaur nesting sites**?

At some sites, many nests were spaced closely together, similar to the colonies or rookeries associated with certain modern sea birds. Evidence shows that certain nesting sites were used over and over again by various dinosaurs. In addition, not only were some dinosaur eggs arranged in a spiral pattern in the nest, but they had a particular vertical orientation, perhaps to minimize breakage. Other nests contained the fossilized bones of young dinosaurs in a wide range of sizes, indicating that the parents cared for the juveniles for an extended period before the young left the nest.

What dinosaur was **named** as a consequence of its **nesting behavior**?

The name of the 26-foot (8-meter) hadrosaur *Maiasaura,* or "good mother lizard," was based on fossil finds in Montana. The nesting colony spanned 2.5 acres and included 40 nests, each containing up to 25 grapefruit-sized eggs. This herbivorous dinosaur showed an advanced social and breeding behavior, including returning to the same nesting sites every breeding season, refurbishing existing nests, placing nests one dinosaur length apart (about 25 to 30 feet [7.6 to 9.1 meters]) so there was room for movement back and forth, incubating eggs using a warm compost layer, and feeding the nest-bound young with vegetation until they were ready to leave.

How have the **fossils** of the ***Oviraptor*** changed our image of this dinosaur?

The dinosaur *Oviraptor,* or egg thief, was previously thought to have been a dinosaur egg consumer, as its fossilized remains were often found near nests. But when scientists uncovered an 80-million-year-old fossil, it showed this bipedal, carnivorous dinosaur (approximately the size of a modern ostrich) was apparently brooding or guarding a nest of 15 large eggs.

The fossil, uncovered in Mongolia's Gobi Desert in the mid–1990s, is the first hard evidence to date showing the behavior of this dinosaur. Other theories on the dinosaur's behavior have been inferred from indirect data. An *Oviraptor* was found lying on its clutch of eggs, with its legs tucked tightly against its body, and the arms turned back to encircle the nest. This is similar to the nesting behavior of modern birds, and it suggests such behaviors may have started long before the advent of wings and feathers.

How did the **dinosaur eggs hatch**?

Similar to modern birds and reptiles, dinosaur young cracked their way out of eggs when they were ready to enter the world. Also similar to these modern species, certain dinosaur species were apparently too small and weak to leave the nest after

Paleontologists can often tell whether the young were independent once they hatched or needed parental care based on the state of the egg fossils found. Some babies may have stayed with their mothers for a long time before they were mature enough to forage for themselves (Big Stock Photo).

hatching. In most cases, the baby dinosaurs probably remained nest-bound for many weeks after they hatched, and had to be fed and tended by the adults. Scientists have deduced this from several nest sites that show the fragments of trampled egg shells and remains that look like regurgitated leaves and berries. Similar to the young of most species, dinosaur hatchlings were no doubt especially susceptible and vulnerable to attacks from predators that hunted around the nesting sites.

How can **parental care traits** of dinosaurs be **extrapolated** from **living animals**?

Of course, the parental care traits of dinosaurs cannot be directly observed because they've been extinct for 65 million years. Therefore, paleontologists have had to resort to observing the closest living relatives of the dinosaurs: birds and crocodiles. They believe that these modern animals probably had many of the same traits, including construction of nests or mounds, nesting together in rookeries, guarding the nest by one or both of the parents, warning and recognition sounds made by the young, and family group cohesiveness during the hatchling stage. In fact, many of these traits have been confirmed by recent fossil findings.

Are there any **examples** of the **dinosaur parenting care methods**?

Yes, there are some examples of parental care methods that scientists have painstakingly pieced together, mostly from known fossil nesting sites around the world. The first example is the *Orodromeus,* an ornithopod that lived in Montana during the Cretaceous. This dinosaur laid its eggs in spirals, with the large ends up and tilted towards the center; the average clutch included 12 eggs. The young hatched with well-developed limb bones and joints, suggesting they could walk almost immediately. This is supported by the low number of crushed eggs in the nest site, indicating the young left quickly. Fossil evidence shows the young stayed in groups, but it is not known for how long.

The *Maiasaura* were also ornithropods in Montana during the Cretaceous. These dinosaurs made nests in shallow holes that were spaced apart from the surrounding nests by about the length of an adult dinosaur. On average, there were 17 eggs per clutch. The fossil evidence to date suggests that the hatchlings had poorly formed limb joints, which meant the young had to stay in the nest for an extended period of time. This conclusion is supported by the numerous fossils of hatchlings found in the nests, along with trampled and crushed eggs. This means that the young *Maiasaura* needed a large amount of parental care and attention, with some estimates having the young staying in the nest for approximately eight to nine months.

Another Cretacious period dinosaur, the *Oviraptor,* was a theropod living in what is now Mongolia. One of the most exciting recent fossil finds is that of an *Oviraptor* in a nesting position, suggesting a brooding behavior similar to modern birds. In contrast to this nurturing attitude, the *Coelophysis*—theropods of the Triassic period found in Arizona—probably ate their young, as seen in many fossil remains.

Were ***Triceratops*** **social** animals?

Similar to today's creatures, not every dinosaur was a social animal. But scientists have discovered that *Triceratops,* which were once thought to be aloof, solitary, shy herbivores, may have actually liked being around others of their kind. The evidence for this was recently uncovered in southeastern Montana in Late Cretaceous rocks. Discovered in the 66-million-year-old rock were the bones of at least three juveniles (they were probably all together when a flood struck the area). Some scientists are not sure, but this behavior may have been for protection, and it may have been more common in juveniles than in adult *Triceratops*.

MATURE DINOSAURS

How **long** did **dinosaurs live**?

Scientists do not know the exact lifespan of the dinosaurs, but they estimate that dinosaurs lived about 75 to maybe as much as 300 years. This educated guess is based on examining the microstructure of dinosaur bones, which indicate that the dinosaurs matured slowly. This is similar to ancestors of the dinosaur, including crocodiles, whose eggs take about 90 days to hatch, and whose lifespans can extend from 70 to 100 years.

What are some of problems with using dinosaur bone growth rings to estimate ages?

Although the analysis of lines of arrested growth (LAG) rings in dinosaur bones has yielded some estimates of growth rates, the technique is not without its problems. It is not known if LAGs truly form in one year. For example, one hadrosaur fossil supposedly had a different number of growth rings in its legs versus its arms. Secondly, growth rates could be size-dependent; in other words, a larger dinosaur might grow slower than a smaller one, just like elephants grow more slowly than mice. Thirdly, some animals may have had indeterminate growth rates like crocodiles. Finally, different species may have different metabolisms, especially when one compares warm- and cold-blooded creatures.

Does the **microstructure** of dinosaur **bones** indicate the **age** of these animals?

Yes, paleontologists have discovered that the bones of dinosaurs have growth rings similar to those found in the trunks of trees that possibly represent each year of a dinosaur's growth. The ring features are very small; to see them, the bones have to be cut into thin sections and examined under a microscope using polarized light. These growth rings are known scientifically as lines of arrested growth (LAG). Although they are assumed to form at the rate of one ring per year, no one truly knows if this is correct. That's because the distances between the growth rings between dinosaurs species vary greatly.

Using this technique, scientists have estimated the age of certain dinosaurs. For example, the bones of a ceratopsian dinosaur, *Psittacosaurus,* indicated that this particular animal was about 10 to 11 years old when it died; a second analyzed dinosaur, a *Troodon,* was three to five years old; the bones of a sauropod called *Bothriospondylus* showed it was 43 years old; a *Massospondylus,* a prosauropod, was 15 years old; and a ceratosaur called *Syntarsus* was seven years old.

What is **another way** to estimate a **dinosaur's life span**?

Another way to estimate a dinosaur's life span is to compare it to the known life spans of living animals, based on body size. In general, larger animals tend to live longer than smaller ones. Using this method, the life span of very large sauropods like *Apatosaurus* and *Diplodocus* was probably on the order of 100 years. Smaller dinosaurs would likely have had shorter life spans.

Do we definitely know the **growth pattern** of dinosaurs?

No, we really don't know for certain what kind of growth pattern the dinosaurs had, and there are several reasons why. First, not all dinosaurs shared the same growth

It might be possible to estimate the age of a dinosaur (here, a *T rex*) by analyzing the growth pattern in its bones (Big Stock Photo).

rate. Different types of dinosaurs probably grew at varying rates, complicating the subject. The second reason for the uncertainty in growth rates has to do with climate. Dinosaurs inhabiting warmer regions of the planet probably grew more rapidly than those living in colder climates. Another reason has to do with the metabolic rate of dinosaurs, which is an area of intense speculation. Warm-blooded vertebrates with higher metabolic rates can grow up to ten times faster than cold-blooded vertebrates. Dinosaurs with potentially higher metabolic rates, such as the small theropods, might have grown faster than the slow moving sauropods, though not larger.

What are the **estimated growth rates** for a **ceratopsian** and **sauropod** dinosaur?

Even though there are concerns with determining growth rates, some scientists have tried to estimate the numbers. In particular, the growth rates were calculated for species where there is fossil evidence from both eggs and adults, using the maximum growth rate of living reptiles as a guide.

For example, an adult *Protoceratops,* a ceratopsian of the Cretaceous period in Mongolia, weighed approximately 390 pounds (177 kilograms); the hatchling weighed approximately 0.95 pounds (0.43 kilograms) (it's assumed the hatchling weight was 90 percent of the egg weight). From this data, the researchers calculated the time the young *Protoceratops* needed to reach adulthood was approximately 26 to 38 years. On the other end of the size scale was an adult *Hypselosaurus,* a sauropod of the Cretaceous period in France. This large dinosaur weighed approximately 5.8 tons (5.3 metric tons) at adulthood; the hatchling weighed approximately 5.2 pounds (2.4 kilograms). The scientists calculated the time need to reach adulthood for this animal at 82 to 188 years.

Are there **problems** with estimating dinosaur **growth rates** based on **modern reptiles**?

Yes, there are problems with using growth rates of modern reptiles as a guide to determine those of dinosaurs. If we look to the modern relatives of the dinosaurs to determine growth patterns, we are again baffled. All these animals also show different types of growth patterns. Reptiles continue to grow as long as they live, though the rate slows with age; this is called indeterminate growth. On the other hand, birds cease growing as they reach adulthood; this is called determinate growth.

Some, if not all, dinosaurs probably had a much different metabolism than modern, cold-blooded reptiles, which would drastically alter the growth rate calculations. Also, the actual growth rate depends on the climate where the dinosaur lived. Animals that live in warmer climates grow faster than those in colder ones.

DINOSAUR QUIRKS

Did any **dinosaurs live** in the colder, **polar regions**?

Yes. Paleontologists believe that while most flourished in tropical or temperate climates, some dinosaurs actually lived in the cold weather regions of the ancient world. Fossils of these polar dinosaurs have been uncovered on the North Slope of Alaska. Others have been found at Dinosaur Cove, at the southeastern tip of Australia, and paleontologists date them to between 110 and 105 million years ago. Although this part of Australia is presently at approximately 39 degrees south latitude, at the time of the polar dinosaurs it was much farther south, lying within the Antarctic Circle. For three months during the winter, the night would have lasted 24 hours, and temperatures were well below zero degrees Fahrenheit (–17 degrees Celsius). The dinosaur fossils uncovered in this area show that the animals were well adapted to these harsh conditions, and that they apparently had keen night vision and may have been warm-blooded. They were generally small animals, ranging in size from about that of chickens to human-sized, with the largest carnivore being about nine feet (three meters) high.

Did dinosaurs **sleep standing up**?

No one really knows the sleeping habits of the dinosaurs. It is not easy to infer such activities based on just the fossil record, as sleeping leaves no definitive physical trace. After all, no one knew sharks slept until just a few decades ago, and sharks are common in today's oceans.

Still, scientists have inferred the sleeping habits of some of the dinosaurs. For example, most of the smaller animals probably slept like modern reptiles, just flopping down on the ground like a crocodile. Others, such as the huge *Tyrannosaurus,* probably had a much harder time sleeping lying down. Once it laid down, it would be difficult to get up using its small arms. Other larger dinosaurs would probably find their enormous weights would get in the way. Thus, the only way larger dinosaurs could sleep was by standing up. It is interesting to note that modern birds (close relatives to the dinosaurs, or maybe even dinosaurs themselves) sleep standing up.

Did dinosaurs **see** in **color** or **black and white**?

Because eyes are soft parts of an animal, they do not survive the fossilization process. Dinosaurs are no different; thus, we have no idea what a dinosaur eye looked like, much less if the animals could see in color or black and white. And it's hard to guess: just look at the diversity of modern animals—and the diversity of eyes, and how and what the various animals see.

Did **dinosaurs** have **binocular vision** similar to humans?

The majority of dinosaurs had monocular vision, with eyes set into the sides of their heads, and little overlap between the right and left fields of view. Thus, they had good peripheral vision, but the binocular vision was modest, similar to the modern alligator. (One of the animals with the best pair of eyes is the modern house cat; they have binocular vision that takes in 130 degrees in front of them, and have peripheral vision that stretches back farther than any other animal.)

But some scientists believe there were exceptions, and some dinosaurs may have had binocular vision similar to a human's depth perception. In particular, predators such as the *Tyrannosaurus* may have been able to see depth, suggesting that the animals were hunters, not scavengers as some paleontologists believe. In addition, over time some carnivores may have evolved facial traits that actually enhanced the animals' ability to see in depth. And some dinosaurs may have developed sight similar to a hawk, a raptor that can see its prey from far away, but whose binocular vision does not kick in until it swoops down to nab its prey. More work is being done by scientists to determine how dinosaurs saw the world by using model dinosaur heads and laser beams to ascertain sight position.

Some hunting dinosaurs probably had binocular vision, just like today's predators. Not all did, though, like this *Coelophysis Bauri*, which was a predator but had eyes better situated for peripheral vision (Big Stock Photo).

How large was a **dinosaur's brain**?

No one really knows the true size of dinosaur brains because, as with all soft parts of dinosaurs, brains did not survive the fossilization process. Therefore, scientists can only infer the size of the animals' brains by examining the brain case: the part of the skull housing the brain. They have found that different dinosaurs had different sized brains. For example, sauropod brains were small in comparison to their body weight, whereas some dinosaurs, such as the *Velociraptor,* had very large brains in comparison to their body weight.

Were **dinosaurs intelligent**?

Unfortunately, there are no dinosaurs around, so there is no way we can determine a dinosaur's intelligence quotient, or IQ. However, we can judge how relatively intelligent dinosaurs were by taking a ratio of brain weight (based on the skull volume) to body weight, then comparing these ratios for various dinosaurs. This ratio is called the encephalization quotient (EQ). Based on this idea, the smartest dinosaurs had the larger brain to body weight ratios than the less intelligent ones.

The following table lists some types of dinosaurs and their EQs. Note that the dromaeosaurids and troodontids were thought to be some of the smartest dinosaurs. The troodontids included the *Troodon,* a carnivore; the dromaeosaurid dinosaurs included the *Velociraptor,* a 6-foot (1.8-meter) carnivore with clawed feet, and sharp, pointed teeth that probably roamed in packs:

Dinosaur Intelligence Based on Encephalization Quotient

Dinosaur	EQ (approximate)
Dromaeosaurids	5.8
Troodontids	5.8
Carnosaurs	1.0–1.9
Ornithopods	0.9–1.5

Dinosaur	EQ (approximate)
Ceratopsians	0.7–0.9
Stegosaurs	0.6
Ankylosaurs	0.55
Sauropods	0.2
Velociraptor	0.2
Sauropodomorphas	0.1

What are the **encephalization quotients (EQs)** for some typical **modern-day mammals**?

The encephalization quotient (EQs) for some typical mammals are based on the same formula as the dinosaurs: The ratio of the brain to body weight. Here are a few EQs of some well-known mammals:

Encephalization Quotients for Modern Mammals

Mammal	Animal Type	EQ
Human	primate	7.4
Bottlenose dolphin	cetacean	5.6
Bluenose dolphin	cetacean	5.31
Chimpanzee	primate	2.5
Rhesus monkey	primate	2.09
Cat	carnivore	1.71
Langur	primate	1.29
Squirrel	rodent	1.10
Rat	rodent	0.40

THE END OF DINOSAURS

THE CRETACEOUS EXTINCTION

What is an **extinction**?

Extinction is the sudden or gradual dying out of a species. There are a multitude of reasons for extinction, ranging from disease, human intervention, climate changes, and natural disasters, such as volcanic eruptions or impacting space bodies. Each one can cause the extinction of one or many species of animals, depending on the severity.

When did the **idea** of **extinction** become **accepted**?

During the seventeenth and eighteenth centuries, scientists knew that fossils were the ancient remains of plants and animals. However, most still thought that these fossils represented known, living species that would shortly be discovered living in some remote, unexplored part of the globe.

This changed radically in the 1750s. Explorers in North America found the remains of what they thought were elephants, but in reality the animals were mastodons and mammoths, which died out more than 10,000 years ago toward the end of the Ice Ages. As these and other fossils from the New World were examined, scientists realized the fossils were actually the remains of extinct species. In 1796, Baron Georges Cuvier of the Museum d'Histoire Naturelle in Paris (the first comparative anatomist) published a series of papers proving these "fossil elephants," and giant mammal bones from other parts of the world, did indeed represent extinct species.

What is the **evidence** for dinosaur and other life form **extinction**?

There are several indicators of dinosaur and other life form extinction in the fossil record. But one of the best ways to determine extinction is by the lack of fossils in

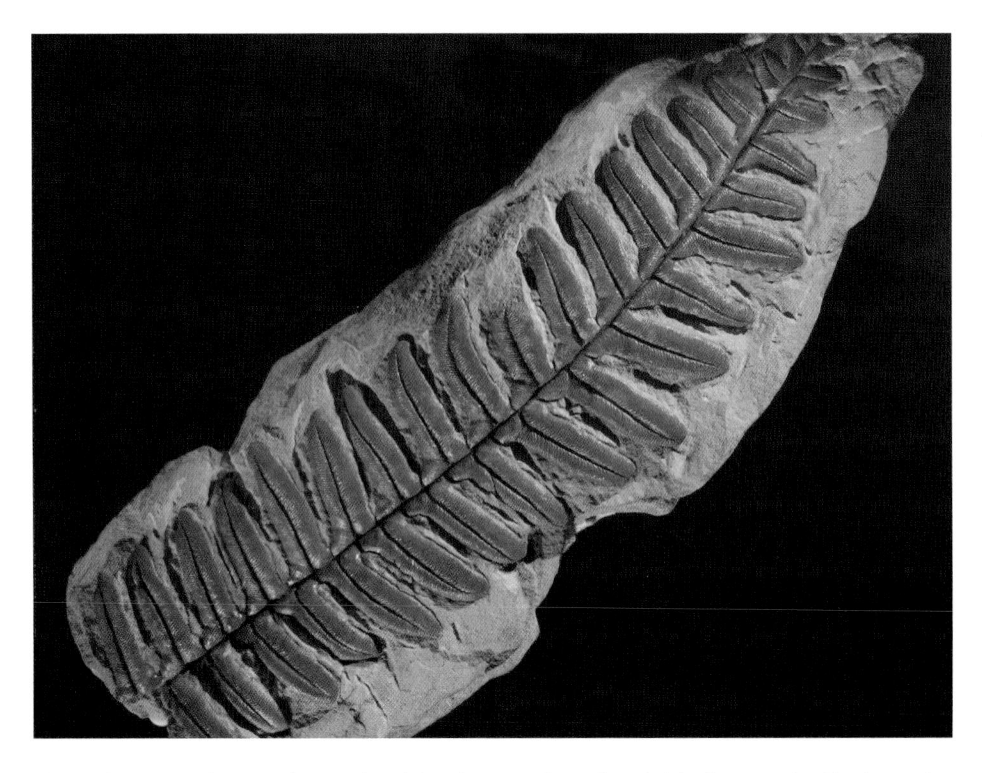

Along with many animal species, a large number of plants became extinct at the end of the Cretaceous. As with other species, ferns died off, as well, but layers of rock filled with fern spores are indicative of how these plants recover quickly from a mass extinction (iStock).

a rock layer. For example, above the top rock layers from the Cretaceous period, there are no known fossils of dinosaurs; and just above the top rock layers from the Permian period, the number of fossils—from animals to plants—greatly diminishes. The reason is logical: the animals that die leave behind fossils; when they become extinct, no more fossils are left behind. Such evidence in rock layers makes it seem as if one minute the organisms were there, and the next they disappeared. In reality, most of the extinctions took place over thousands of years.

What is the **"fern spike"**?

The "fern spike" is a layer of rock filled with fern spores; it occurs after a major cataclysmic or localized extinction. Scientists believe that after a major mass extinction, most plants would be wiped out. The first plants to recover are the ferns, which spread their spores into the air: thus, the "fern spike." This evidence in rock layers is often used to determine the line between the time before and after a massive global or local extinction.

What were the **major extinctions** during Earth's long history?

Around five major extinctions have occurred over Earth's history. Some of the extinctions greatly affected the animals and plants on land, while other extinctions mainly occurred in the oceans. Most of the major extinctions are based on the fossil record, and usually indicate a time when a large percent of the plants and animals living on Earth went extinct, usually for unknown reasons. The following lists some of the known major extinctions on the geologic time scale:

Major Extinctions

Time Period	Date (millions of years ago)	Percent of Species Extinct (approximate)
Cambrian-Ordovician	438	84%
Devonian-Carboniferous	360	82%
Permian-Triassic	250	97%
Triassic-Jurassic	208	76%
Cretaceous-Tertiary	65	85%

How many species of **dinosaurs** were living at the end of the **Cretaceous period**?

No one knows the exact number of dinosaur species living before the end of the Cretaceous period, as we do not have a complete dinosaur fossil record. The nature of fossilization—or that not all animals were in the right place at the right time in order to become a fossil—and the process of erosion have wiped away much of the evidence over time. Although scientists know there were probably many more dinosaurs, only a few species were still alive toward the end of the Cretaceous period, and most of them lived on the North American continent.

How many **other organisms** became **extinct** at the end of the **Cretaceous period**?

No one knows the exact number of other organisms—land animals, marine animals, and plants—that became extinct at the end of the Cretaceous period. But based on the fossil record, scientists believe that about 85 percent of all species on Earth went extinct at the end of the Cretaceous period, thought to be the second largest extinction (the Permian was first).

What were some **other organisms** besides dinosaurs that **became extinct** at the end of the Cretaceous period?

There are many other animals that became extinct at the end of the Cretaceous period. They include the flying pterosaurs and many types of ammonids, marine reptiles, redist bivalves, brachiopods, mollusks, and fish.

What **animals didn't** experience as many **extinctions**?

There were several animal groups that did not experience as many extinctions among species. They included most mammals, birds, turtles, crocodiles, lizards, snakes, and amphibians.

What **plants** became **extinct** at the end of the Cretaceous period?

Many species of plants became extinct at the end of the Cretaceous period, but few among the ferns and seed-producing plants. In North America alone, almost 60 percent of plant species became extinct.

Why did certain **plants and animals survive** through the end of the **Cretaceous period**?

Until a definite reason for the extinction is determined, it is difficult to answer such a question as why some animals and plants survived. Apparently, some species were not affected by the occurrence. In fact, nocturnal mammals—probably through luck or inborn tolerance to harsh environmental conditions—largely survived. They quickly exploited all the new places available to them, and soon dominated the planet. Eventually, over millions of years, humans evolved from those species.

Can scientists see the **Cretaceous boundary** in **rock**?

Yes, there are rock deposits all over the world that show the characteristic mineral signature that represents the Cretaceous boundary. Because they are found all over the globe, scientists know that no species was unaffected by whatever events occurred to create the mass extinction at the end of the Cretaceous period.

Why aren't **huge amounts** of **dinosaur bones** found in late Cretaceous period rock layers?

This is another mystery surrounding the whole question of dinosaur extinction at the end of the Cretaceous. If indeed the dinosaurs were suddenly killed off by a catastrophe, there should be a thick layer of bones—or a "bone spike"—in the fossil record. However, no such bone spike has been found to date at the boundary between the Cretaceous and Tertiary periods. Equally puzzling, few dinosaur bones have been found within a foot or so below this boundary.

One possible theory to explain these "missing" dinosaur bones (if they truly are missing) involves acid rain. Models have shown that one consequence of a large asteroid impact would be highly acidic rainfall over the planet. This acidic water could have dissolved most of the dinosaur bones lying on the surface, and would have also penetrated below the surface into the upper soil zones. Combined with bacteria, the water would become even more acidic, dissolving any bones found there. Only already fossilized bones would have resisted the acidic water. Since the fossilization process takes a very long time, none of the more recent dinosaur bones would have been spared.

This theory—and it is only a theory at present—neatly explains why there are so few dinosaur bones found in rock below the boundary, and none at the boundary itself. Supporting evidence comes from the boundary layer: in many places around the world, a relatively thin layer of clay exists that could have formed from the erosion of rocks due to acid rain.

What "**bone spike**" was found in **Late Cretaceous** period rock layers?

A Late Cretaceous "bone spike," the thick layer of bones found in the fossil record around the time of a catastrophic extinction, has not been found until recently. A large bed of fossil fish bones from this time has been discovered on Seymour Island, Antarctica, covering more than 31 square miles (50 square kilometers). Although there is a possibility that the fish were killed off by volcanic activity, climate change, or some other environmental cause, their bones lie immediately above the iridium rich layer that marks the end of the Cretaceous period. In other words, it's highly likely that the fish were victims of the catastrophic extinction that also affected the dinosaurs.

THEORIES ABOUT DINOSAUR EXTINCTION

How long ago did the **dinosaurs** disappear?

Based on the dinosaur fossil record, dinosaurs died out about 65 million years ago. More recently, many scientists point out that not all dinosaurs disappeared, citing birds as direct descendants of the dinosaur. In fact, a recent study shows that birds may have survived the Cretaceous-Tertiary extinction because they had larger and more complex brains than many other contemporary animals, including the dinosaurs. Such physical features probably made survival—and competition—that much easier for the birds, allowing them to better adapt to the environmental changes associated with the mass extinction.

Were the **dinosaurs inferior** because they became extinct?

The fact that dinosaurs became extinct has been used as proof of their inferiority and that they were evolutionary failures. But it is unlikely that anyone can claim the inferiority or superiority of a species. Some species, such as the *Lingula* clams, have been around for 500 million years; the dinosaurs lived for about 150 million years; and the earliest hominids appeared only about three million years ago, with modern humans (*Homo sapiens sapiens*) appearing only about 90,000 years ago. If, as many paleontologists believe, birds are really proven to be dinosaurs, then we may one day say that dinosaurs have existed for almost 200 million years.

What are the **two general theories** as to how fast the dinosaurs went **extinct**?

The two general theories are catastrophism and gradualism. Catastrophism is the rapid change in conditions found on Earth, such as changes in the atmosphere, that led to the death of most species of dinosaurs. Gradualism says that the dinosaurs died out slowly, over a period of many hundreds of thousands or millions of years. This may be due, for example, to changes of climate caused by continental drift. Some scientists think both theories are correct, with all the changes coming together at the end of the Cretaceous period and leading to the extinction of the dinosaurs.

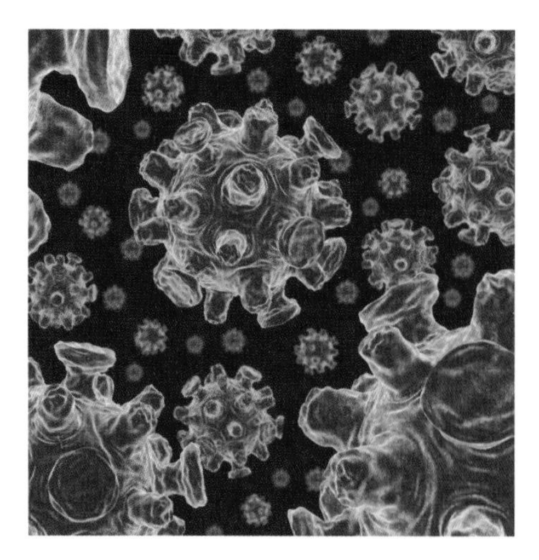

Even with all of our modern technology and medicine, diseases such as the flu (caused by the influenza virus) claim thousands of human lives each year. So it doesn't take a huge leap of the imagination to speculate that a virus or single strain of bacteria could have wiped out the dinosaurs (iStock).

What is the **"disease theory"** of dinosaur extinction?

The "disease theory" of dinosaur extinction (which falls under the theory of gradualism) states that dinosaurs eventually died out because of disease. Some say biological changes, brought about by changes in their evolution, made the animals less competitive with other organisms, including mammals that had just started appearing. Others say a major disease—anything from rickets to constipation—wiped out the dinosaurs, with some dinosaur bones definitely showing signs of these diseases over time. Another idea is that overpopulation led to the spread of major diseases among certain species of dinosaurs, and eventually to them all.

Could the **atmosphere** have caused the **extinction of the dinosaurs**?

Some scientists have proposed an atmospheric cause for the extinction of the dinosaurs. In particular, they point to a reduction in the amount of oxygen present in the atmosphere during the Cretaceous period. Researchers recently measured microscopic air bubbles trapped in amber. The study showed that the amount of oxygen present two million years before the end of the Cretaceous period was approximately 35 percent; just after the end of the Cretaceous, the amount was down to 28 percent.

Would a **lower oxygen level** affect **dinosaurs**?

Yes, with a lower oxygen level, the dinosaurs could have experienced extreme respiratory stress. This is similar to what humans encounter when living or working at extremely high altitudes without supplemental oxygen.

Today's atmospheric oxygen level is 21 percent, but modern animals have the proper physiology to live in this atmosphere. Dinosaurs first evolved when the atmosphere was oxygen rich, and they had the right physiology for this situation. Studies of an *Apatosaurus* skeleton show a limited capacity to breath, with relatively small nostrils and probably no diaphragm. This was fine as long as the atmosphere was rich in oxygen, but it was inadequate when the oxygen levels fell.

Under this scenario, dinosaurs may have experienced three different events leading their gradual extinction. First was a cooling climate toward the end of the Cretaceous period. Second, the oxygen levels might have fallen, making it extreme-

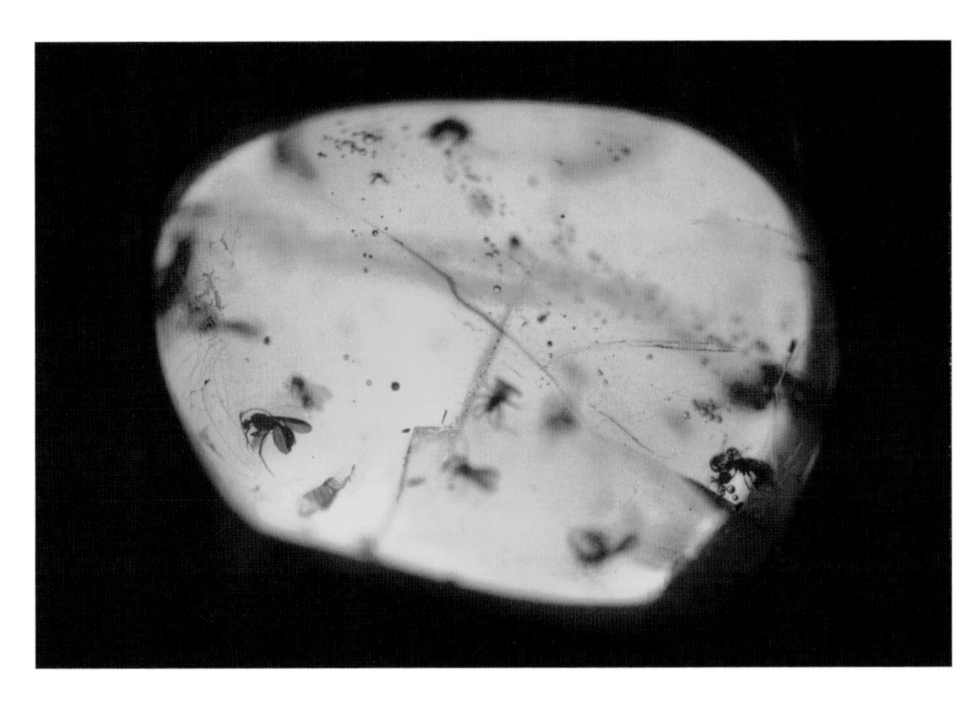

Amber, which is fossilized tree sap, has been found to sometimes preserve fossil evidence, such as ancient insects, and even bubbles of oxygen, which may be analyzed for clues about the prehistoric atmosphere (iStock).

ly difficult to breath. This probably led to a reduction in dinosaur diversity, shrinking the number of genera from 35 about 10 million years before the end of the Cretaceous period to only 12 at the end. At that point, a third and final catastrophic event may have occurred: An asteroid impact or impacts, and/or volcanic activity, which finally pushed the remaining dinosaurs to extinction.

What is the **theory** concerning **radiation killing** off the dinosaurs?

According to recent findings, dinosaurs and other animals may have gone extinct because of epidemics of cancer. And although highly debated, some scientists believe these epidemics may have been caused by a massive burst of neutrinos—or weakly interacting elementary particles from a star with no electric charge and apparently no effective mass—from dying stars in our galaxy.

What could have caused **lower oxygen levels** at the end of the **Cretaceous period**?

The reduction in oxygen levels in Earth's atmosphere toward the end of the Cretaceous period may have been caused by volcanic activity. Volcanic activity could have increased, modifying the relative amounts of atmospheric gases, such as carbon dioxide and oxygen. In turn, this would influence the evolution of life on the planet. This is known as the "Pelé hypothesis," after the Polynesian goddess of volcanoes.

Many people today are concerned that global warming—evidenced by such phenomena as melting ice caps—will lead to species extinctions, including, possibly, our own. Scientists speculate that environmental changes may have also led to the dinosaurs' extinction (iStock).

What is the **"mammal theory"** of dinosaur extinction?

The "mammal theory" of dinosaur extinction (which falls under the theory of gradualism) states that dinosaurs were slowly wiped out by mammals, animals that only appeared around the end of the Mesozoic era (the end of the Cretaceous period). The mammals could have eaten many of the dinosaurs eggs (thus, the dinosaurs had difficulty reproducing); or the mammals could have taken over territory from the dinosaurs. On a smaller scale, such events happen today, especially when an introduced species takes over another species by eating the native organism's young, or by taking over the territory and eating the available food supply.

What is the **"climate theory"** of dinosaur extinction?

The "climate theory" of dinosaur extinction is one that many scientists believe is more feasible than most other theories. One of the theories of gradualism is that the movement of continents over millions of years brought about changes in Earth's climate, including the changing of oceanic currents; spreading of deserts; drying up of inland seas; shifts in Earth's axis, orbit, or magnetic field; spreading of polar ice caps; and the increase in volcanic eruptions. The resulting slow climate changes (from either one, several, or all of the above events) caused the gradual decline of the dinosaurs. They could not evolve quickly enough to compensate for the changes.

What is the **"poison plant theory"** of dinosaur extinction?

The "poison plant" theory of dinosaur extinction involves the development of angiosperms (flowering plants), a new type of plant that first flourished during the Cretaceous period. Some of the plants were no doubt poisonous to dinosaurs, and the plants probably developed protective toxins to prevent being eaten. The more

What is the "human theory" of dinosaur extinction?

There is no such thing. The idea of humans living at the same time as dinosaurs seems to have been propagated by the B-movies of the 1950s and 1960s in which dinosaurs attacked humans. In reality, mammals did evolve about the same time as the dinosaurs, but the mammals were not human-like. The dinosaurs died out about 65 million years ago; the first hominids appeared about three to four million years ago; *Homo sapiens sapiens* (or modern humans) did not appear until about 90,000 years ago.

prolific plant-eating dinosaurs may have died out as the plants became more toxic to these animals; in turn, the carnivores had fewer plant-eating dinosaurs to eat.

But this theory is too simplistic. There were many plants in the world, including varieties that were non-poisonous. In addition, the idea does not explain the mass extinction of marine organisms—animals that had nothing to do with flowering plants on land—at the end of the Cretaceous period.

Did **dinosaurs** get blown away by **hurricanes**?

No one really knows, but several researchers think this may have been possible. These scientists studied huge hurricanes called hypercanes, monster storms that grow much larger than modern hurricanes, especially if the ocean water is greatly warmed. They believe a large impacting meteorite struck or a major volcano erupted in shallow ocean waters, causing the ocean water temperatures to rise, doubling the temperatures we currently find in the tropics. This increase in water temperature could have created immense hypercanes. In turn, the storms could have carried water vapor, ice crystals, and dust high into the atmosphere, blocking sunlight and destroying the protective ozone layer that shields animals from the ultraviolet radiation of the Sun. The effect could have devastated the dinosaurs. Scientists acknowledge the idea is a little far-fetched, but not impossible.

THE IMPACT THEORY

What is the **"impact theory"** of dinosaur extinction?

The impact theory is one of the newest ideas in the catastrophic camp of dinosaur extinction. This theory states that a large object (or objects), such as an asteroid or comet, collided with Earth, resulting in a large impact crater, giant waves in the oceans that smashed onto land at heights of two to three miles (3.2 to 4.8 kilometers), and radical, rapid changes in Earth's weather, temperature, amount of sunshine, and climate.

What are **asteroids** and **comets**?

An asteroid is a large rocky body found in outer space. They range from boulder size all the way up to about 600 miles (965 kilometers) wide; they generally are classified as either carbonaceous, stony, or metal. The majority of asteroids are found along the plane, or ecliptic, of the solar system. The larger ones are sometimes called minor planets because of their propensity to orbit along the same plane as the major planets. Most of the asteroids revolve around the Sun in a tight band between the orbits of Mars and Jupiter, called the asteroid belt. Italian astronomer Giuseppe Piazzi (1746–1826) discovered the first asteroid, Ceres, in 1801.

Comets are a collection of dust, gases, and ice that orbit the Sun. Once described as "dirty snowballs," many comets are now thought to be more like "mudballs," most carrying more dust than ice. In general, comets are composed of carbon dioxide, frozen water, methane, ammonia, and materials such as silicates and organic compounds.

Where are **asteroids** and **comets found** in our solar system?

The majority of the asteroids stay within the asteroid belt, a band of chunks of rock between the orbits of Mars and Jupiter. Short-period comets, or those with orbits from a few to 200 years, are thought to have originated in the Kuiper Belt, a fat disk of comet-like objects beyond the orbit of Neptune and Pluto. Long-term comets, or those that travel into the solar system every few thousand years (or may never return at all), are thought to originate in the Oort Cloud, a theoretical cloud of comets proposed by Dutch astronomer Jan Oort (1909–1992). The cloud surrounds the solar system about 100,000 astronomical units from the Sun (one astronomical unit is equal to about 93 million miles [149,637,000 kilometers], the average distance between Earth and the Sun).

What are **near-Earth objects**?

Over hundreds of thousands of years, because of the gravitational pull of the planets or other space objects, an asteroid or comet may stray from the belt. Such space objects—mostly asteroids—that come close to Earth or cross Earth's orbit are called near-Earth asteroids (the ones that cross Earth's path are also called Earth-crossing asteroids). It is known that in the past some near-Earth asteroids struck Earth, creating impact craters on the planet's surface. Meteor Crater in Arizona is a good example of an impact crater formed by an asteroid. There also seems to be an association between some of the larger Earth impact craters and the extinction of a large number of species during the planet's long history.

Scientists currently believe there are about 2,000 near-Earth objects, (mostly asteroids or burned-out comets) larger than a half mile (one kilometer) in diameter that revolve around the Sun in short-period orbits. These objects can occasionally intersect the orbit of Earth; but most of the time, we pass right by each other or we are far from each other when the object crosses Earth's orbit. The only problem is that scientists believe that only about 7 to 10 percent of this estimated pop-

One of the theories about dinosaur extinction that has gained favor is that a large asteroid struck Earth, dramatically changing the environment. Today, astronomers are tracking near-Earth objects in order to prevent a similar fate from befalling us (iStock).

ulation has been discovered. Although hampered by funding cuts, scientists continue to search the sky for these potential hazards. Therefore, we do not know the actual orbits of most of these near-Earth objects.

Who **first hypothesized** the **impact scenario**?

In 1980, the American physicist Luis Alvarez (1911–1988) proposed that a large asteroid or comet hit Earth about 65 million years ago. His son, geologist Walter Alvarez (1940–), discovered a high concentration of iridium (an element associated with extraterrestrial impacts) at the Cretaceous-Tertiary (nicknamed "K/T") boundary in Italy. Because of this find, and the realization that dinosaurs and many other species died out at the end of the Cretaceous, Luis and Walter Alvarez, along with colleagues Frank Asaro and Helen Michel, proposed that the extinctions at the K/T boundary were caused by the impact of a large space object. The iridium anomaly has since been found in over 50 K/T boundary sites around the world.

What are **impact craters** and are there any on **Earth**?

An impact crater is a large hole in the surface of a planet that is most often caused by a collision with a large space object, such as a comet or an asteroid. All planets and most satellites have impact craters, and even asteroids have impact craters. The Moon is our most obvious example of impacts on a planetary body, as the surface is dotted with hundreds of craters.

To date, scientists have identified over 150 impact craters on Earth. Most have been found on the surface; fewer than a dozen or so are buried or found deep in the oceans. There were probably many more craters, but erosion—from wind, water, or the movement of the continental plates—has erased any evidence of their existence. In addition, there may be many more craters under the thick, vegetative growth of the jungles, in high mountains, or buried deep under sediment on land or in the oceans. The largest known crater, the Vredefort crater in South Africa, is also one of the oldest, with an age of over two billion years. Another large impact crater, the Sudbury in Canada, is a major source of certain metals. Craters on the planet Mars dwarf Earth's craters. The largest impact crater (also called a basin) on the red planet is Hellas Planitia, measuring 1,243 miles (2,000 kilometers) in diameter.

Large Impact Craters

Name	Location	Diameter (miles/kilometers)
Vredefort	South Africa	186/300
Sudbury	Ontario, Canada	155/250
Chicxulub*	Yucatan, Mexico	105/170
Manicouagan	Quebec, Canada	62/100
Popigai	Russia	62/100
Acraman	South Australia, Australia	56/90
Chesapeake Bay	Virginia, USA	53/85
Puchezh-Katunki	Russia	50/80
Morokweng	South Africa	44/70
Kara	Russia	40/65
Beaverhead	Montana, USA	37/60

*Crater thought to be associated, or at least partially associated, with the extinction of the dinosaurs.

What **evidence** supports the **impact theory**?

The most compelling evidence for this theory is the Chicxulub crater near the Yucatan Peninsula in Mexico, an impact crater that was discovered by geologists in 1992. This almost 180-mile- (110-kilometer-) wide crater is thought to be the result of a 6-mile- (10-kilometer-) diameter asteroid. The crater was created approximately 64.98 million years ago, which would be the right time frame for the extinction of the dinosaurs. The crater is buried, and was actually found in the 1960s during a subsurface survey taken by an oil company. It took years before a geologist looking at the data noticed the circularity of the feature and brought the impact crater to the attention of the scientific community.

Why was the **impact** at **Chicxulub** so **devastating**?

The impact at Chicxulub around 65 million years ago would have been devastating enough, but the effects were worse because the asteroid traveled at such a shallow angle. The approximately 6-mile- (10-kilometer-) wide object struck the Yucatan at an angle of approximately 20 degrees above the southeast horizon. If the impact had

Because of erosion and other active geological processes on Earth, impact craters such as Barringer Meteorite Crater in Arizona, are uncommon. The theory that a devastating impact from an asteroid killed off the dinosaurs at first seemed like an unlikely explanation to many scientists (iStock).

occurred straight down (perpendicular to Earth) most of the energy would have been directed into the planet's interior. But the shallow impact angle of the Chicxulub object meant that debris—in the form of vaporized and molten rock—was scattered forward towards the northwest. This instantly destroyed all living organisms, including dinosaurs, over western North America. The material forced into the upper atmosphere by this angled impact would have cooled the climate over a period of months, effectively killing off the remaining dinosaurs.

How **many impact craters** are associated with dinosaur extinction?

Currently, there is only one known Earth impact crater associated with dinosaur extinction: the Chicxulub crater in Mexico. In addition, tiny glass fragments from the impact ejecta (the rock and soil that sprayed from the crater when it formed) were found in 1990 in the Caribbean Ocean on the island of Haiti. The ejecta debris appears to fall in line with the Chicxulub crater. Another crater, the Manson structure in Iowa, was once thought to have formed at the end of the Cretaceous, but subsequent studies show it is not the correct age. Two impact crater sites in Belize and Mexico, about 290 miles (480 kilometers) and 140 miles (230 kilometers) from the Chicxulub crater, respectively are thought to be from the ejecta of Chicxulub—the material thrown up from the crater that landed nearby.

How long **after the asteroid impact** did the dinosaurs become **extinct**?

No one truly knows how long after the asteroid impact the dinosaurs became extinct, but several studies in 2009 point to a possible number: around 300,000

If dinosaurs did become extinct from impacts, how did they die?

After the impact, scientists theorize that several events occurred. First, huge amounts of dust and debris were thrown high into the atmosphere. This dust was then carried all around the world by the upper atmospheric winds, filtering or blocking out the sunlight. At the same time, heat from the blast may have created firestorms—huge forest fires that added smoke, ash, and particles to the already dust-filled atmosphere. If the dust-filled winds did not kill the animals, the lack of sunlight would have killed off plants, creating a serious crisis. The animals feeding off of the vegetation would die, and in a domino-like effect the rest of the organisms in the food chain would die off, including the dinosaurs.

years after the impact. For example, scientists uncovered marine sediments in Mexico that seem to contain at least 52 species of various animals, creatures that were found in younger rock. Another scientist claimed to have unearthed dinosaur bones in the American Southwest that date almost a half million years after the Yucatan impact. If this is true, scientists don't believe that the dinosaurs actually survived in areas around the impact, but instead the creatures that survived in relatively unscathed areas later moved and recolonized in areas closer to the impact.

Has there been any evidence of a **recent killer asteroid**?

Yes, scientists believe they have found evidence of a more recent killer asteroid. This huge rock smashed into southeastern Argentina about 3.3 million years ago, possibly killing off 36 species of animals because of the local climate change. Neither the asteroid itself nor a crater was found, but glassy fragments (called escoria) were dug up from the soil. These fragments indicate intense heating that could only have come from a space object striking Earth.

Unlike the asteroid that supposedly struck at the end of the Cretaceous period, the Argentina hit did not affect the global climate. But the strike apparently did change the local climate, cooling down the temperatures. This killed off many animals in the region, including giant armadillos, ground sloths, and a large-beaked carnivorous bird.

OTHER EXTINCTION THEORIES

Did changes in **Earth's environment** lead to dinosaur **extinction**?

Although the predominate dinosaur extinction theory is that an asteroid impact dramatically changed Earth's climate, there are some paleontologists who think

Once very common in the world's oceans, ammonites suffered a huge die-off and eventually became extinct around the time of the Chicxulub impact (iStock).

otherwise. These scientists believe that changes in Earth's environment played a much more important role in the extinction of life forms on Earth and that the impact 65 million years ago just killed off the remaining species.

Recent studies of the fossil record have examined changes in a wide range of plant and animal species' populations toward the end of the Cretaceous period. Only single-celled marine life showed a sudden decline at the end of this period. Other species declined gradually, with a few groups showing no change. And there was no evidence for a catastrophic mass extinction in the fossil record.

Instead, the majority of extinctions may have occurred gradually due to environmental changes such as falling sea levels and volcanic eruptions. The sea level apparently dropped by approximately 328 feet (100 meters) during this time. Also, debris from volcanic eruptions in India may have been thrown into the atmosphere. Both could have contributed to a gradual extinction of many flora and fauna species.

What is a **problem** with the **environmental theory** of dinosaur extinction?

The main problem with the environmental theory of dinosaur and other life form extinctions lies in the conflicting evidence in the fossil record. For example, evidence for the environmental theory includes ammonite (hard-shelled relatives of squid) fossil records. One record shows they were in decline for approximately 11 million years before finally going extinct, but other fossil evidence indicates the impact at Chicxulub suddenly killed off from one-half to three-quarters of ammonite species along the coasts of France and Spain.

Unfortunately, the fossil record of Cretaceous dinosaurs is very limited for the last 10 million years of that period. In fact, the only good dinosaur records from this time are in western North America—and this evidence is limited to the last two million years of the Cretaceous period. Thus, this poor, discontinuous dinosaur fossil record may be the real cause for the apparent reduction in their diversity, and not environmental causes.

What are **killer cosmic clouds**?

Killer cosmic clouds are large areas in outer space, probably bigger than our entire solar system, that have much higher concentrations of hydrogen than normal. For the past five million years, our planet has been traveling in a relatively empty, typical region of space, with a density of less than one particle (mostly hydrogen) per cubic inch. Killer clouds are found where new stars are being formed and have much higher densities of hydrogen on the order of hundreds of particles per cubic inch.

Some scientists believe that a killer cloud could collapse the solar system's heliosphere—a bubble of space produced by the solar wind that partially protects our planet (and the other planets and satellites in our solar system) from cosmic rays. Cosmic rays are high-speed particles from outer space that constantly hit the heliosphere, but most are deflected by this shield. This is good because exposure to the powerful radiation from these rays could fry a human being. If the heliosphere around our planet collapsed from the introduction of a cosmic cloud, much higher levels of cosmic radiation would strike Earth, dramatically altering life, though scientists are not sure how much or in what ways.

Could a **killer cosmic cloud** have caused the **extinction** of the **dinosaurs**?

If supercomputer models are correct, then higher concentrations of hydrogen could have formed a wall and caused the heliosphere around Earth to collapse. This could have allowed more cosmic radiation to penetrate to Earth's surface, resulting in changes to the flora and fauna. If this did occur, such an increase in cosmic rays, with elevated levels of radiation, could have directly killed the dinosaurs. Another scenario is that the rays negatively affected the vegetation eaten by the herbivorous dinosaurs and other animals. As carnivorous dinosaurs no longer had any prey, they also would have died off.

Is there any **proof** that Earth has encountered a **cosmic cloud** in the past?

No, there is no proof of encountering a cosmic cloud. Some scientists believe that, from time to time, our planet could have encountered smaller, less dense clouds of hydrogen, known as the Local Fluff. These less-devastating encounters, according to the models, would have only weakened the heliosphere, with slight increases in cosmic rays hitting Earth.

One of the known side effects of cosmic rays striking Earth is the production of the rare metal beryllium. An increase of this metal could be proof that we had encountered one of these relatively benign clouds. Ice cores taken from the South Pole do indeed show an increase in beryllium levels at approximately 35,000 and 60,000 years ago, leading scientists to speculate that contacts were made with Local Fluff. What are the effects of these minor encounters? Scientists speculate that the effect of Earth coming into contact with the Local Fluff might have produced anything from an ice age to an increased greenhouse effect.

Could a killer cosmic cloud have caused the mass extinction on Earth 65 million years ago? The high concentration of hydrogen in space clouds like this one may have temporarily collapsed Earth's heliosphere, according to one theory (NASA).

What is the **true story** of **dinosaur extinction**?

Many scientists believe that dinosaurs became extinct not due to one reason, but rather from a combination of reasons, most of which are examined above. In addition, some scientists believe that dinosaurs were already gradually declining when catastrophe occurred 65 million years ago. There is a chance that they would have become extinct anyway—with or without a catastrophe happening.

AFTERMATH OF EXTINCTION

What is the **Cretaceous/Tertiary boundary** and what does it indicate about the **extinction** at the end of the **Cretaceous** period?

The mass extinction at the end of the Cretaceous period is evident in the Cretaceous/Tertiary (K/T) boundary, a thin layer of rock found in various areas around the world—all dominated by a large amount of iridium, a chemical element found in abundance in comets and asteroids but rare on Earth. It was this K/T boundary that first clued scientists in to the fact that there was a mass extinction about 65 million years ago. According to this layer and other evidence in the fossil record, mass extinctions of animals and plants—in both oceans and on land—occurred around the same time.

What groups **survived** the **extinction** at the end of the Cretaceous period?

The survivors include most land plants and land animals—insects, snails, frogs, salamanders, crocodiles, lizards, snakes, turtles, and certain mammals. Most marine invertebrates also survived, including starfishes, sea urchins, mollusks, arthropods, and most fishes.

What was a **common characteristic** of the surviving land animals?

The land animals that survived were all small in stature, such as mammals, frogs, and snakes. The larger animals, such as the dinosaurs, did not survive this extinction. In fact, some scientists estimate that no animal heavier than 55 pounds (25 kilograms) survived on land.

What is the Cretaceous–Paleogene boundary?

The Cretaceous-Paleogene (or K-Pg) boundary is what some scientists are calling the Cretaceous/Tertiary boundary, the Paleogene being the period from about 65 million to 23 million years ago.

What groups **did not survive** the **extinction** at the end of the Cretaceous period?

The groups that did not survive the massive extinction include the dinosaurs, pterosaurs, and some families of birds and marsupial mammals on land. In the oceans, mosasaurs, plesiosaurs, some families of teleost fishes, ammonites, belemnites, rudist, trigoniid, and inoceramid bivalves became extinct, as well as over half the oceans' various plankton groups. Some groups appear to have vanished completely at the end of the Cretaceous period, whereas others were already gradually diminishing in diversity in the last 10 million years of the Cretaceous period.

How did the **end** of the **Cretaceous** period **rank** among Earth's extinctions?

The mass extinction at the end of the Cretaceous period was the second largest in geologic history; around 76 percent of all species disappeared. The largest mass extinction was at the end of the Permian period, about 250 million years ago, in which about 90 to 97 percent of all species disappeared.

What types of **plants** did not disappear at the end of the **Cretaceous period**?

Although many species of plants disappeared at the end of the Cretaceous period, many ferns and seed-producing plants continued to survive into modern time.

What percentage of **marine animals** went **extinct** at the end of the **Cretaceous period**?

The marine biota was hit very hard by the Cretaceous extinctions. About 15 percent of all marine families died out, as well as perhaps 80 to 90 percent of all species. Here are some examples of the approximate percentage of certain groups that went extinct:

- Ammonites—100%
- Marine reptiles—93%
- Planktonic foraminifera—83%
- Sponges—69%
- Corals—65%
- Sea urchins—54%
- Ostracodes—50%

All of the flying pterosaur species, like this *Quetzalcoatlus*, became extinct at the end of the Cretaceous (Big Stock Photo).

What percentage of **land animals** went **extinct** at the end of the **Cretaceous period**?

At the end of the Cretaceous period, many land animals became extinct. For example, about 25 percent of all land animal families disappeared. About 56 percent of the reptiles in general died out, and 100 percent of all non-avian (bird) dinosaurs and pterosaurs became extinct.

Why did **certain animals survive** and other animals did not?

No one is really sure why certain animals died out and others did not. In some ways, the animal extinctions at the end of the Cretaceous period were very selective.

Are **mammals** surviving **relatives** of the **dinosaurs**?

No, mammals are not the surviving relatives of the dinosaurs. The earliest mammals were descendants from certain types of reptiles, but they are not in the same line as the dinosaurs.

What **mammals** lived at the **end** of the **Cretaceous** period?

Mammals had been around for millions of years before the end of the Cretaceous period; in fact, the first group of true mammals, the morganucodontids, evolved in the late Triassic period. They were a successful group of animals for about 150 million years before the dinosaurs became extinct.

By the end of the Cretaceous period, some mammals had developed many innovations vital to their survival. Many stopped laying eggs and were able to deliver live young. Various mammal species eventually grew specialized teeth for a variety of tasks, such as cutting, gnawing, and grinding, for the better processing of food. They developed better ways to compete for food, such as having more energy in proportion to their size, or adapting to changing diets by becoming omnivores (plant- and meat-eating animals).

The therian mammals (marsupials and placentals) became the apparent heirs to the land the dinosaurs and other organisms left behind. Some mammal subgroups had already disappeared before the demise of the dinosaurs; others made it through the end of the Cretaceous period; and some even survive to this day.

Why did the **mammals** come to **dominate** in the **Cenozoic era**?

Mammals came to dominate the Cenozoic era (our present time) because there was "suddenly" little competition. The larger predatory reptiles had disappeared, so the mammals were able to quickly fill the available ecological niches.

Are **mammals** the most **abundant animals** on **modern Earth**?

No, mammals are not the most abundant animals—in terms of species or individuals—on our planet. There are many more kinds of fish, reptiles, and birds; and there are even more invertebrate species on Earth, including insects and mollusks.

What are the **closest living relatives** to the **dinosaurs**?

The closest living relatives to the dinosaurs are thought to be certain modern reptiles and birds.

What is a **reptile**?

Modern reptiles include the alligators and crocodiles, turtles, lizards, and snakes. They all have several typical characteristics: They have a protective covering of scales or plates, they lay eggs, they are cold-blooded, they breathe with lungs instead of gills, and they have five clawed toes on each foot (with exceptions, of course, such as snakes; and alligators only have four toes on their back feet, but five on their front feet).

When did **modern reptiles evolve**?

The earliest turtles evolved during the Triassic period, but they probably could not withdraw into their shells like modern turtles. Lizards and snakes have poor fossil records. This is probably due to the animals' tendencies to live in dry uplands, far from the areas that are most likely to produce fossils (most animal bones survive if they are quickly buried with sediment, such as along riverbanks). It is thought that lizards appeared in the Late Triassic; the earliest remains of snakes are found in the Late Cretaceous (in North America and Patagonia, South America).

Modern snakes belong to the group of animals called reptiles; all reptiles have scales, lay eggs, and are cold-blooded (iStock).

How **well** did **reptiles** fare at the **end** of the **Cretaceous** period?

Today's reptiles, like many animals at the end of the Cretaceous period, are almost like the survivors of a shipwreck. After the Cretaceous period, most reptiles were wiped out. About 6,000 reptiles species exist today, fewer in numbers and much smaller in size than their ancestors, but greater in diversity.

What reptile **species survived** past the **end** of the **Cretaceous** period?

One remarkable group of reptiles—also close relatives of the dinosaurs—are the crocodiles. They appear to have evolved from archosaurian ancestors during the Late Triassic, but unlike most of their contemporaries they survived through to the present day. They are also remarkable for the fact that these moderate- to large-sized, semiaquatic predators have remained relatively unchanged since the Triassic period.

What are the **two** types of **modern crocodiles**?

There are two types of modern crocodiles found in tropical and subtropical environments. The gavialids are found in India; they eat fish and have slender snouts. The crocodylids are found almost worldwide and consist of crocodiles and alligators. They have long bodies and powerful tails used for swimming or defense. Their limbs allow the animals to maneuver and steer in the water; on land, they use their limbs to walk with a slow gait, with their bellies held high off the ground. These animals survive off a wide variety of food, including fish, large vertebrates, and carrion.

What is the **difference** between an **alligator** and a **crocodile**?

Although there is some overlap of habitats with alligators and crocodiles, it is rare to see both together. The best way to tell the difference between the two animals is by checking the size and head: Crocodiles are slightly smaller and less bulky than an alligator. In addition, the crocodile has a larger, narrower snout, with a pair of enlarged teeth in the lower jaw that fit into a "notch" on each side of the snout. The

Both crocodillians, an alligator (top) has a number of physical characteristics that are different from a crocodile (bottom), including the shape of the snout, arrangement of teeth, and body size (iStock).

alligator has a broader snout, and all the teeth in its upper jaw overlap with those in the lower jaw.

How might **dinosaurs** have **evolved** if they had **not gone extinct** 65 million years ago?

Dale Russell, curator of fossil vertebrates at the Canadian Museum of Nature in Ottawa, Canada, believes that dinosaurs were evolving toward more human-like features toward the end of the Cretaceous period, including the development of a larger brain, forward-focused eyes, and bipedalism. Extrapolating these tendencies, he "evolved" a dinosaur called a *Troodon*. He came up with a bipedal creature he called a Dinosauroid, which, though reptilian in many ways, including its extremities and somewhat scaly skin, also looked very humanoid.

FROM DINOSAURS TO BIRDS

What are **birds**?

Birds are members of the animal kingdom; they have their own class, known as Aves. (A possible origin of the word "bird" is thought to be the Old English *brid,*

While you definitely wouldn't have heard "Polly want a cracker!" during the Cretaceous period, evidence has been found suggesting an ancient ancestor of the parrot may have existed during that time (iStock).

which originally referred to the young of animals.) Birds are vertebrate animals, warm-blooded, reproduce by laying eggs, and have feathers and beaks. They have four limbs, with the front two limbs modified into wings.

How are **birds classified**?

Birds have a class all their own called Aves. Within this broad class is a subclass called the Neornithes. This grouping includes the approximately 10,000 species of living birds. Another classification term for these birds is Neoaves, a basal division of birds that includes 95 percent of all living birds. In turn, this grouping is further divided into four suborders based on the palate anatomy of the birds. (There also are some different classification systems for birds, including one that divides the Neornithes only into the Palaeognathae and the Neognathae; for purposes of this discussion, we will use the four suborder classifications.) The four suborders are the Palaeognathae, which are divided into two subgroups, the ratites (such as ostriches, rheas, emus, and other large, flightless birds) and tinamous (such as the South American tinamous); Impennes (penguins); the Odontognathae (fossil birds); and Neognathae (all other living birds, from hummingbirds to plovers).

Were any **birds around** when **dinosaurs** roamed Earth?

Yes. Scientists believe that the ancient ancestors of several modern birds were around during the latter part of the Cretaceous period. Originally, scientists

What did Archaeopteryx hear?

Many scientists believe the *Archaeopteryx* was both the earliest bird and a dinosaur. Now scientists may have another piece of evidence to connect birds and dinosaurs: a study in 2009 discovered that the *Archaeopteryx* had a hearing range similar to that of the modern emu. This conclusion was based on the size of the cochlear duct, part of the creature's inner ear that contains sensory tissue. Analysis of the ducts on emus and a fossil *Archaeopteryx* revealed that the dinosaur had an average hearing range of 2,000 hertz (humans hear from 20 to 20,000 hertz), which is a number similar to the modern emu. (The emu, by the way, has one of the most limited hearing ranges among modern birds.)

thought that only aquatic birds, such as ancestors of the loons, seabirds, and albatrosses populated Earth during prehistoric times. More recently, ancient bones resembling a modern parrot have been discovered. As more bones have been found and confirmed, scientists have concluded that certain birds also flew when the dinosaurs roamed Earth about 65 to 70 million years ago.

Did some early scientists believe **birds** were **related to dinosaurs**?

Yes, some early scientists believed there was a similarity between birds and reptiles, an idea that was noted as far back as 400 years ago. But the relationship connection did not come to the forefront of science until the mid-nineteenth century, especially after the discovery of bird-like bones from a German rock quarry.

Who first published **papers** noting the **resemblance** of **birds to dinosaurs**?

In 1867, paleontologists Edward Drinker Cope (1840–1897) and Othniel Charles Marsh (1831–1899) were the first to publish papers noting the resemblance of birds to dinosaurs.

What is ***Archaeopteryx lithographica***?

Archaeopteryx lithographica (literally "ancient wing from lithographic limestone") fossils are dated at around 150 million years old and are some of the world's most famous fossils. Thought by many modern paleontologists to represent the oldest bird yet discovered, it was a small bird about the size of a crow. The first fossil of an *Archaeopteryx lithographica* was found in 1855. The fossil remains were found in sedimentary rock from the Late Jurassic period in the Solnhofen quarries in southern Germany. Interestingly enough, the fossil would not be recognized as a bird until 1970.

Who first **proposed** the idea that **birds** were **descendants of dinosaurs**?

English naturalist Thomas Henry Huxley (1825–1895), an authority on bird evolution and a champion of Charles Darwin's theory of evolution, first noted characteristics shared by dinosaurs and birds. Huxley observed that a particular species of chicken raised in his area (called a Dorking fowl) had leg bones similar to those of theropod dinosaurs. He also cited as evidence the fossil of a bird-like dinosaur, *Archaeopteryx lithographica,* discovered in 1855 in Germany. Between 1868 and 1869, using a method of anatomical comparison resembling modern cladistic analysis, Huxley decided birds had descended from dinosaurs. Years later, the *Archaeopteryx lithographica* represented the transition between dinosaurs and birds, and it was thought of as proof that birds descended from dinosaurs.

English naturalist Thomas Henry Huxley was the first to theorize that birds may have descended from dinosaurs (iStock).

Have **other skeletons** of *Archaeopteryx* been found?

Yes, other *Archaeopteryx* fossils have been discovered: a total of 11, so far. An almost complete skeleton was found in 1861 and was referred to as the "London specimen." The fossil was the basis for a continuing debate between supporters and detractors of Charles Darwin's then-newly published theory of evolution.

What are some of the ***Archaeopteryx*** fossils **found** to date?

There have been 11 actual specimens and many feathers found of the *Archaeopteryx* to date. Here is a list of nine discoveries made so far:

The Haarlem specimen was found near Reidenburg in 1855, five years before the feather was discovered. Because it was not known to be a fossil of an early bird, it was classified as a *Pterodactylus crassipes,* or pterodactyl (not even a dinosaur); in 1970, John Ostrom examined the fossil and found evidence of feathers, and thus, its true identity. It is currently at the Teylers Museum in Haarlem, the Netherlands.

The London specimen was found in 1861 near Langenaltheim. It was eventually bought by the British Museum of Natural History (under the instruction of Richard Owen) from Carl Haberlein.

Can scientists tell the color of fossil feathers?

Until recently, scientists were only guessing about the colors of fossil feathers. But researchers may have a clue: traces of organic material found in some fossil feathers are actual remnants of pigments that once gave the birds their color. These were once thought to be carbon traces from bacteria, but are now known to be fossilized melanosomes, the organelles that contain melanin pigment. Since it appears that melanin can be preserved in certain fossils, scientists may now have a way to reliably predict the original colors of feathered dinosaurs, including the *Archaeopteryx* and its relatives.

The Berlin specimen was found in 1876 or 1877 near Blumenberg, and sported a complete head (although it was badly crushed). It is housed at the Humboldt Museum für Naturkunde.

The Eichstatt specimen was found in 1951 or 1955, depending on sources, and is the smallest of all the *Archaeopteryx lithographica,* measuring about two-thirds the size of the other specimens; it also has the most well-preserved head found so far. It had a different tooth structure and its shoulder bones are not ossified as much as the other specimens, making many scientists believe this animal is an example of a different genus. Other scientists believe that the fossil represents a juvenile *Archaeopteryx lithographica,* or a species from an area with different food, thus the different structures. It is currently at the Jura Museum in Eichstätt, Germany.

The Maxburg specimen was found in 1958 near the same place as the London specimen in Langenaltheim. The fossil represents the animal's torso only and is the only specimen to be privately owned. It was found by Eduard Opitsch, who died in 1991; after his death, the specimen was found to be missing, and is thought to have been secretly sold. Thus, the whereabouts of this specimen remains a mystery today.

The Solnhofen specimen was found in the 1960s near Eichstatt and was at first thought to be a *Compsognathus*. After preparing the specimen in the lab, scientists noticed that its arms were too long for its body size; they also found feathers, and so the creature joined the list of *Archaeopteryx lithographica*. It is currently at the Bürgermeister-Müller Museum in Solnhofen.

The Munich specimen (formerly the *Solnhofen-Aktien-Verein* specimen) of an *Archaeopteryx* has a small ossified sternum and feather impressions. Interestingly enough, it was found in 1991 and described in 1993; it is classified by some scientists as a new species: *Archaeopteryx bavarica*. It is currently at the Paläontologisches Museum in Munich.

The Bürgermeister-Müller specimen was uncovered in 1997. This fragmentary fossil is currently at the Bürgermeister-Müller Museum.

Finally, the *Thermopolis specimen,* long in a private collection, was discovered in Germany and described in 2005.

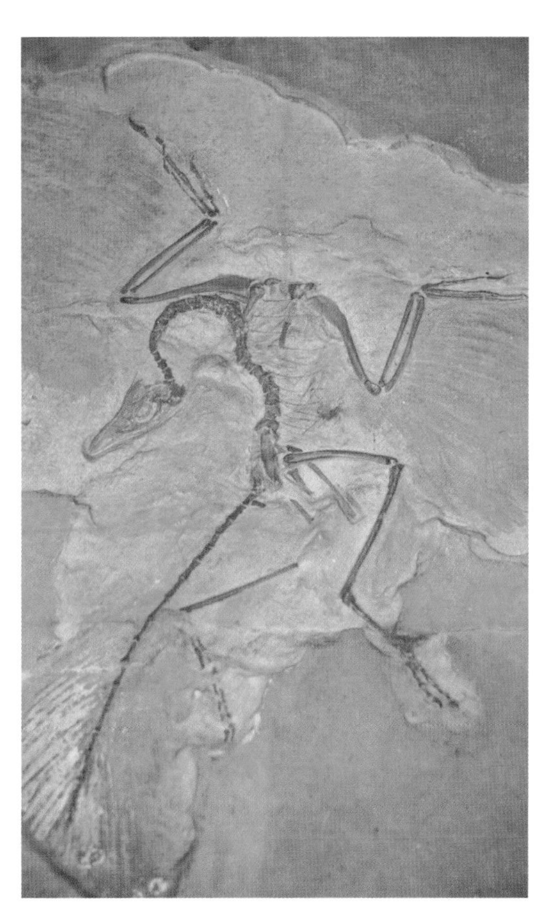

Widely believed to be one of the first ancestors of birds, the Archaeopteryx first appeared in the Jurassic period (iStock).

What has been discovered about ***Archaeopteryx lithographica*** fossil **feathers**?

Besides the actual specimens of the *Archaeopteryx lithographica,* many fossil feathers have been found. One of the first was discovered in 1860 near Solnhofen and described in 1861. It was also a surprise to scientists, not because it was old, but because of the feather's exquisite detail. So far, researchers believe that the *Archaeopteryx* had feathers almost all over its body, except the upper neck and head. This may be because the feathers were not easily preserved in the fossil, or there was some physical reason for the absence of feathers on specific parts of the body.

Did some scientists believe ***Archaeopteryx lithographica*** was a **transition** between **dinosaurs** and **modern birds**?

Yes, some scientists once believed *Archaeopteryx* was the transition between dinosaurs and modern birds, mainly because the fossilized skeletons exhibited a mixture of dinosaur- and bird-like features. The dinosaur (or reptilian) characteristics include such features as bony tails, teeth, and claws on the fingers; the bird-like characteristics include such features as feathers, wishbones, and beaks. Today, however, many scientists believe *Archaeopteryx lithographica* may have just been a link in dinosaur progression, eventually evolving into modern birds.

Does everyone believe ***Archaeopteryx*** was a link to dinosaurs?

No, not everyone believes *Archaeopteryx* was a direct link to the dinosaurs. Some scientists believe birds and dinosaurs evolved separately from a common reptilian ancestor, but so far, no one has yet found acceptable fossil evidence to support or disprove this idea.

Who more recently revived the idea that birds were the descendants of dinosaurs?

The idea of birds as a kind of dinosaur was revived in 1969 by paleontologist John Ostrom. He brought out the idea that dinosaurs may have been warm-blooded, thus, more active and similar to birds. Robert Bakker's article in *Scientific American* the next year pursued the same ideas. At this time, scientists also began to delve deeply into dinosaur physiology (cells and tissues), noticing the physiological similarities and differences between the animals and other species such as birds. It is these studies, in combination with skeletal evidence, that scientists hope will lead to the correct answer concerning bird lineage.

What was the bird-like animal named ***Confuciusornis sanctus***?

The *Confuciusornis sanctus,* discovered in the northeast province of Liaoning, China, was a pigeon-sized, flying creature that may have appeared after the *Archaeopteryx*. (Its actual age is still debated, but it may be about 140 million years old, which is about 10 million years after versus the *Archaeopteryx*.) This animal had a horny, toothless beak. Before it was discovered, scientists thought toothless beaks did not appear until the Late Cretaceous period, about 70 million years ago. This creature also had feathers along its legs, making this (so far) the earliest known record of contour feathers on any animal.

DINOSAURS ALL AROUND US

What are the major camps in the **dinosaur-bird evolution debate**?

There are several camps of paleontologists in the dinosaur-bird evolution debate. One group believes birds descended from certain dinosaurs about 60 million years ago. Another camp believes proto-birds evolved separately from dinosaurs about 200 million years ago. And there is another group that has emerged: scientists who believe that birds are actually dinosaurs. Right now, there are not enough fossils to come to a definite conclusion, and all sides have good arguments. However, with the advent of DNA sequencing, scientists may one day have the answer.

What is **cladistic analysis**?

Cladistic analysis is a method used to determine an organism's family tree. The older system of classification, developed by Carolus Linnaeus (Carl von Linné, 1707–1778) in the eighteenth century, categorizes plants and animals by organisms' overall similar characteristics. Cladistic analysis uses specific characteristics, such as wrist bones, and relates them to previous and following generations, thus tracing the evolution of these structures. The more characteristics previous and following generations share, the more likely they are related. (A cladogram represents a diagram of all the clades, or groups of organisms.)

Cladistic analysis is not easy. Scientists must study the minute details of early animal fossils, noting the tiniest differences in bones and joints. Each different characteristic is assigned a code and added to a computer database. The computer then sorts the information, producing what looks like a "family tree," linking together past and modern animals by these detailed characteristics.

What does **cladistic analysis** tell us about **birds** and **dinosaurs**?

According to cladistic analysis, birds share some 132 characteristics with dinosaurs. Some scientists believe this hard evidence indicates that birds are, indeed, a kind of dinosaur. But many scientists still disagree.

How long ago did **modern birds originate**?

Scientists are still debating this question. Many researchers have suggested, based on fossils—and cladistic analysis—that modern birds arose around 60 million years ago. But more recent fossil discoveries have caused the debate to become heated, and some researchers suggest modern birds may have evolved more than 100 million years ago. The problem comes with how scientists look at bird evolution: some use the fossil records and others use genetic data, both of which yield conflicting results. New techniques, such as DNA sequencing, may help scientists resolve when and how the world's close to 10,000 bird species originated.

What recent **genome study** may rewrite the **evolutionary history of birds**?

The debate about the origin of birds—and their connections to dinosaurs—has reached into the molecular lab. For many years, the Early Bird Assembling the Tree-of-Life Research Project, centered at The Field Museum in Chicago, has been examining DNA sequences from all major living bird groups. So far, more than 32 kilobases of nuclear DNA sequences from 19 different locations on the DNA sequencing have been checked for each of 169 bird species. This massive undertaking has already revealed some birds "secrets." For example, distinctive bird lifestyles, such as being nocturnal or raptorial, may have evolved several times. One example is the colorful, daytime hummingbirds that evolved from the darker-colored, nocturnal nightjars. As the studies continue, researchers already know that not only will the names of dozens of birds have to be changed, but so will many biology, paleontology, and birdwatchers' field guides.

What overall **characteristics** are seemingly **shared** by **dinosaurs** and **modern birds**?

Not all dinosaur characteristics are similar to those of modern birds, but there are many similarities. For example, some dinosaurs had features such as bony tails,

Birds share over 130 physical traits with dinosaurs, and some species look remarkably like dinosaurs with feathers—for example, birds in the ratite family, including ostriches, emus, rheas, and this cassowary from New Zealand. The cassowary even sports a head crest reminiscent of animals like the *Dilaphosaurus* (Big Stock Photo).

claws on the fingers, beaks, and on some, feathers.

Feathers are only one major characteristic that links dinosaurs to birds, including such feathered dinosaur fossils as the *Archaeopteryx lithographica,* and specimens found in other parts of the world, such as the *Sinosauropteryx prima* from northeast China. All these fossils exhibit feather-like impressions in the sedimentary rock in which they were found.

Other characteristic links between dinosaurs and birds involve certain skeletal similarities. For example, fossils of the dinosaur *Deinonychus* have many bird-like characteristics: the large head was balanced on a slender, almost bird-like neck; and the chest was short, with the arms folding inward in a resting position, similar to the wings of a bird at rest. The creature's foot had a huge claw on the second toe. In many ways, the feet of the *Deinonychus* resembled an enlarged representation of a bird's foot.

How did **feathers evolve**?

Scientists really don't know yet how feathers evolved. Some scientists believe feathers probably began as modified scales. These changed scales were not intended for flight, but rather for insulation to preserve the reptile's body heat. Eventually, they evolved into feathers. Other scientists believe feathers evolved from scutes similar to the thick scales on the top of a bird's foot. Analysis shows that the bird scutes, scuttelae, claw sheathes, beak sheathes, and scales around a bird's eyes have the same chemical composition as feathers, and they are controlled by the same genes. Crocodiles, a sister group of the dinosaurs, also have scutes, with a similar (but not identical) chemical composition to bird scutes. Scientists also know that dinosaurs had scutes, too, but they don't know the composition. Of course, to confuse matters even more, some scientists speculate that scutes evolved from feathers.

Have any dinosaur proteins withstood millions of years of burial?

Yes. Recently, researchers at the North Carolina State University found fragments of proteins in the 80 million-year-old remains of a *Brachylophosaurus canadensis,* a duck-billed dinosaur. This was an amazing discovery, since few scientists believed a protein could last more than 100,000 years. Researchers nicknamed it a "dino-protein," and found that the dinosaur's protein fragment was similar to that found in chickens and ostriches. This helped support the bird–dinosaur connection, one of the most controversial theories in dinosaur studies today.

What **dinosaurs** are thought to have evolved into **birds**?

Paleontologists who believe birds evolved from dinosaurs think that the most likely bird ancestors were the small, carnivorous theropods. At least one fossil finding seems to indicate that dromaeosaurs, a subgroup of the theropods, eventually branched into many lines, including birds. This subgroup line also included such dinosaur species as *Velociraptor, Deinonychus,* and *Utahraptor*.

Do scientists believe the ***Microraptor*** could **fly**?

The recently discovered *Microraptor gui*—a feathered dinosaur found in northeastern China that lived about 125 million years ago—was only 30 inches (77 centimeters) long. These creatures had not one, but two sets of wings, one on top of the other, making them look like a kind of early animal "biplane." Because of this, scientists believe that the *Microraptor* could not fly, but would glide from tree to tree, or even hop and glide from place to place. Unlike many birds today, the *Microraptor* probably could not land on the ground after jumping from a tree. Based on its structure, its wings did not act like a parachute, so they would have crashed to the ground.

What is the **latest theory** stating **birds** are truly **dinosaurs**?

One recent theory goes one step further than "birds are descendants of dinosaurs." Some paleontologists now think modern birds are *truly* dinosaurs. This theory states that birds are dinosaurs that evolved certain bird-like characteristics, such as light weight, agility, wings, feathers, and beaks. These specialized characteristics enabled the animals to somehow survive the extinction at the end of the Mesozoic era and continue evolving into modern birds.

Do all paleontologists believe that **birds and dinosaurs** are related?

No, not all paleontologists believe that birds are dinosaurs or even that birds evolved directly from dinosaurs. Many suggest that birds and dinosaurs descended from a common, older ancestor, and developed many superficial similarities over

This *Scelidosaurus harrisonii* had a birdlike beak instead of teeth. Beaks in some dinosaurs are one reason some paleontologists see parallels between dinosaurs and modern birds (Big Stock Photo).

millions of years due to convergent evolution; because they both developed body designs for bipedal motion, they eventually started to resemble each other.

There are four major reasons why some scientists think there is not a direct dinosaur-bird link: the timing problem, body size differences, skeletal variations, and the mismatch in finger evolution. First, with regard to timing, there is some fossil evidence that bird-like dinosaurs evolved 30 to 80 million years after the *Archaeopteryx lithographica,* which seems to put the cart before the horse, or the bird before the bird-like dinosaur. This is the opposite of what you would expect if birds descended from dinosaurs.

Second, some scientists suggest it was nearly impossible for theropods to give rise to birds because of differences in body size. They point out that these carnivorous dinosaurs were relatively large, ground-dwelling animals, with heavy balancing tails and short forelimbs, not the sort of body that could evolve into a lightweight, flying creature.

Third, although birds and dinosaurs have skeletons that appear similar at a macroscopic level, there are really many variations on a smaller scale. The teeth of theropods were curved and serrated, but early birds had straight, unserrated, peg-like teeth; dinosaurs had a major lower jaw joint that early birds did not; the bone girdle of each animal was very different; and birds have a reversed rear toe for perching, while no dinosaur had a reversed toe.

Fourth, there is a discrepancy between the fingers of dinosaurs and birds. Dinosaurs developed hands with three digits, or fingers, labeled one, two, and three, and corresponding to the thumb, index, and middle fingers of humans. The fourth and fifth digits, corresponding to the ring and little fingers, remained tiny vestigial bumps, which have been found on early dinosaur skeletons. However, as recent studies of embryos have shown, birds developed hands with fingers two, three, and four, corresponding to the index, middle, and ring fingers of humans. Fingers one and five, corresponding to the thumb and little finger, were lost. Some paleontologists wonder how a bird hand with fingers two, three, and four could have evolved from a dinosaur hand with fingers one, two and three. And they say the answer is: It's impossible.

THE SEARCH FOR THE MISSING LINK

What fossil evidence possibly shows that **not all dinosaurs became extinct**?

Several recent fossil discoveries are used as proof that not all dinosaurs became extinct, but are the animals we call birds. The fossil findings of *Caudipteryx zoui* and *Protarchaeopteryx robust* from China are seen by some paleontologists as proof that birds not only descended from dinosaurs, but are dinosaurs. The remains of these two 120 million-year-old species were found in Liaoning Province in northeast China. Two fossils were *Protarchaeopteryx robust,* while a third was a *Caudipteryx zoui*.

Protarchaeopteryx robust fossils indicate this animal was a relative of *Archaeopteryx lithographica*; it was turkey-sized and covered with down- and quill-like feathers. The *Caudipteryx zoui* fossil remains show features of a theropod dinosaur, but it was also covered with down- and quill-like feathers. The wing feathers were not swept-back (as needed for flight), but were symmetrical in shape.

Both animals had the shape of swift, long-legged runners and, based on their skeletal characteristics, were more closely related to dinosaurs than birds. Even though they had feathers, neither appeared capable of flight. This seems to indicate that certain dinosaurs developed feathers for reasons other than flight. These feathered dinosaurs did not go extinct at the end of the Cretaceous period, and some scientists believe they continued to evolve into modern birds.

What fossil discovery indicated a **link** between **birds** and **dromaeosaurs**?

A fossil that may link birds and dromaeosaurids is named *Rahona ostromi,* or "Ostrom's menace from the clouds," in honor of American paleontologist John Ostrom (1928–2005). This fossil was found in 1995 off the eastern coast of Africa on the island of Madagascar. Dating of the fossil indicates that the animal lived about 65 to 70 million years ago, or the Late Cretaceous period. The fossil remains show a primitive bird about the size of a modern raven, with a 2-foot (0.6-meter) wingspan. The also has tiny bumps along the wing bones, indicating where the flight feathers were attached.

Many paleontologists are convinced that *Rahona ostromi* had feathers and could fly. The first toe of each foot points backwards, indicating the animal had a perching foot similar to that in modern birds. But it also retained many dinosaur-like features, the most interesting of which is on the second toe of each foot. *Rahona ostromi* had a sickle-shaped, killing claw on its second toes, a feature found in the subgroup of theropod dinosaurs known as dromaeosaurids.

What is the bird-like animal named ***Shuvuuia deserti***?

The 80 million-year-old fossil remains of *Shuvuuia deserti*—derived from the Mongolian word for "bird" and the Latin for "desert"—were found in the Gobi Desert of Mongolia. This was the first skull found from an animal belonging to a family called the *Alvarezsauridae,* which some scientists believe represents an advanced stage in the transition from dinosaurs to birds. *Shuvuuia deserti* was flightless, turkey-sized, walked on two legs, had a long tail and neck, and had short forearms ending in a single, blunt claw. Later specimens have shown a reduced second and third finger.

Although it was more advanced than the earlier *Archaeopteryx lithographica, Shuvuuia deserti* did not look like a stereotypical modern bird, leading paleontologists to conclude that birds in the Late Cretaceous period were as diverse as they are today. Although many of these primitive species during the Cretaceous were quite different, they did have some unique characteristics found in modern birds. In the case of *Shuvuuia deserti,* the similarity was prokinesis—the up and down, independent movement of the snout that allowed the mouth to open wide.

What is the bird-like animal named ***Unenlagia comahuensis***?

Unenlagia comahuensis, or the "half-bird from northwest Patagonia," was a dinosaur living approximately 90 million years ago. Similar to many other recent reptile discoveries, the animal had bird-like characteristics that interest scientists.

Some scientists believe that *Unenlagia comahuensis* is the actual missing link between birds and dinosaurs. Although apparently too big to fly, the *Unenlagia comahuensis* had several characteristics in common with birds, including arms that could flap like wings and a bird-like pelvis. In particular, the 5-foot- (1.5-meter-) tall dinosaur had a shoulder blade (scapula) that shared some characteristics with birds; the forearm socket in this bone pointed out toward the side rather than down and back, as seen in other primitive, bird-like dinosuars such as the *Deinonychus*. In addition, the *Unenlagia*'s triangular pelvis looks like a cross between those in the *Deinonychus* and *Archaeopteryx*.

What is the bird-like animal named ***Sinosauropteryx prima***?

Fossils of a *Sinosauropteryx prima,* found in China, are of a theropod dinosaur with what appears to be feathers. At first, scientists believed feathers ran down the dinosaur's back. Upon closer study, it is now thought the animal actually had feather-like structures. The reason for this arrangement is debated. Some scientists believe the structures were used for movement; other scientists believe they were probably protofeathers, an early step toward the evolution of bird feathers. The *Sinosauropteryx* is one of the "Chinese feathered" dinosaurs and is a more recent

It is believed that feathers were originally used by dinosaurs like this *Gigantoraptor erlianensis* for ornamentation or possibly insulation, and not for flying (Big Stock Photo).

evolutionary development than the *Archaeopteryx*. Some scientists believe they are one of the most primitive coelurasaurs, as seen from their skeletal features.

What do the ***Sinosauropteryx*** and other similar dinosaurs tell us about the **evolution of feathers**?

The protofeathers like the ones on the *Sinosauropteryx* appear to be covered with hollow filaments. These may have been the forerunners of bird feathers; or, as some scientists believe, feathers used for insulation, not the precursors to flight. Dinosaurs like the *Sinosauropteryx,* other similar dinosaurs, and fossil feathers from the same period, give scientists a better idea about the evolution of feathers. In particular, it is thought that feathers evolved rapidly in terms of geologic time. Feathers possessed properties that were highly advantageous to survival; they gave the animals an aerodynamic body and provided insulation for warmth, two features needed to survive a tough, predator-ridden world.

What was the **original purpose** of **feathers** on dinosaurs?

Based on the recent fossil discoveries of feathered, flightless dinosaurs, some paleontologists now think feathers were originally developed for insulation or ornamentation purposes.

Were **birds** the **only animals** that sported **feathers**?

Because geologic time takes in so many millions of years, some scientists believe there were other early animals that sported some type of feather or protofeather. Many researchers believe that modern birds are direct descendants of two-legged, meat-eating dinosaurs, and since there was a wide range of these prehistoric

What baby bird fossil seems to link dinosaurs and birds?

A 135-million-year-old baby bird was found in the Pyrenees of northern Spain. It had wings, feathers, and tiny holes in its immature bones just like modern birds, but it also had a head and neck similar to carnivorous dinosaurs, with sharp teeth and powerful neck muscles for chewing. This fossil still is highly debated regarding its possible connection to dinosaurs and birds.

theropods, there's a chance other dinosaurs had feathers, either as juveniles or all their lives. Some species kept their feathers (ancestors of birds, for example), while others did not. The reason for the lack of evidence may be because feathers were not preserved, or maybe such fossils have yet to be discovered.

How did early **feathered animals** eventually **develop flight**?

There are several theories on how early feathered animals developed flight. One theory states that gliding creatures increased their surface area from the body outward, but flying animals increased it away from their center of gravity, giving them lift and more maneuverability to escape predators. Another theory states that wing motions evolved from the hunting techniques of the coelurosaurs. The skeletal structure of these dinosaurs allowed them to swing their arms down from above and behind their shoulders, allowing for easier grabbing of prey. Both these theories imply that birds developed flight from the ground up, not from the trees down, which is also known as the cursorial theory.

Other paleontologists believe that flight developed in animals that lived in trees (the arboreal, or tree, theory); jumping out of these trees provided enough acceleration to generate lift. Still others believe that some animals ran and flapped their wings, increasing their angle of attack on the down stroke and generating enough lift to fly, especially if they were running downhill.

What carnivorous **dinosaur** fossils had traces of a **bird-like beak**?

In the Red Deer River Badlands of Canada, the remains of a carnivorous ornithomimid dinosaur were found that included traces of keratin around the front of the animal's skull. Keratin, the material found in hair and fingernails, is also found in the beaks of birds. This was also the first carnivorous dinosaur with evidence of a beak, and it showed that dinosaurs could make the transition from teeth to a beak-like structure.

What does the internal structure of ***Scipionyx samniticus*** reveal about the **ancestry** of **birds**?

Studies of the 113-million-year-old *Scipionyx samniticus* fossil's interior structure and unique metabolism has led some scientists to say that none of the known groups of dinosaurs could have been the ancestors of birds. This is because birds have an entirely different lung structure than the *Scipionyx samniticus* (a small coelurosaurid theropod), and a much different metabolism. However, this is based on the findings from one fossil, one of very few to have internal parts preserved (parts of the windpipe, intestines, muscles, and liver). These results will have to be carefully integrated with other findings, such as the presence of feathers and common skeletal features, to come to any useful conclusions.

Hitchcock presented a paper describing the footprints in stone found in the Connecticut Valley. He collected over 20,000 of these fossil footprints during his life-time, and organized the world's largest collection at Amherst College. He believed giant birds made the tracks, not lizards or reptiles.

The first dinosaur track, similar to this one, was officially described in the early nineteenth century and was discovered near Holyoke, Massachusetts (iStock).

What was the **first dinosaur track** to be **described**?

The first dinosaur track to be described anywhere was a three-toed track from the east bank of the Connecticut River near Holyoke, Massachusetts. This large footprint cast was originally named *Orinithichnites giganteus,* but was later renamed *Eubrontes giganteus*. A lithograph of this track was incorporated into William Buckland's *Bridgewater Treatise* in 1836.

What early United States **expedition** mentioned a **giant leg bone** that was probably from a dinosaur?

The early United States expedition was that of army officers Meriwether Lewis (1774–1809) and William Clark (1770–1838)—the two explorers sent out by President Thomas Jefferson (1743–1826) to find a northwest passage to the Pacific Coast. In 1806 the explorers recorded finding a giant leg bone near Billings, Montana; it was almost certainly that of a dinosaur.

When were the **first fossils** found in the **Western Hemisphere** accurately identified as belonging to a **dinosaur**?

In 1856, American paleontologist Joseph Leidy (1823–1891), professor of anatomy at the University of Pennsylvania, accurately identified several fossil bones as being those of dinosaurs. These fossil remains were among the first to be collected in the western United States by an official geological survey team in 1855. The bones were found in what is now Montana. The remains were mostly fossil teeth and were subsequently shown to be from *Trachodon* and *Deinodon* dinosaurs.

Who first suggested that some **dinosaurs** were **bipedal**?

Originally, dinosaurs were thought of as either giant, sprawling lizards, or bulky, quadrupedal reptiles with some mammal-like features. But, in 1858, American paleontologist Joseph Leidy (1823–1891) described an almost complete skeleton discovered by W.P. Foulke in Haddonfield, New Jersey. The fossil skeleton, named a

Are there dinosaur reconstructions in New York City's Central Park?

No, but there used to be dinosaur reconstructions in New York City's Central Park. In 1868, sculptor Benjamin Waterhouse Hawkins (1807–1889), who made the sculptures at Crystal Park in England, was commissioned to make several dinosaur reconstructions in Central Park. According to reports, Hawkins tangled politically with "Boss Tweed" and his gang; Tweed smashed the models and threw them in the park's lake. All that is left are the drawings of the dinosaurs. There are other reports, too, concerning these dinosaur reconstructions. Some say the dinosaurs are buried under the park, but the original story remains the most accepted.

Hadrosaurus, was more complete than any yet discovered. It also indicated that the dinosaur was bipedal—a radical notion for its time.

What was the **first mounted dinosaur cast** reflecting a **bipedal stance** in the United States?

In the 1860s, at the Philadelphia Academy of Sciences, a mounted cast skeleton of a *Hadrosaurus* was the first to reflect a bipedal stance. Named and described by Joseph Leidy (1823–1891) in 1858, this *Hadrosaurus* was the first cast skeleton in North America to be free-mounted. Working with Leidy on this project were paleontologist Edward Drinker Cope (1840–1897) and Benjamin Waterhouse Hawkins (1807–1889), who was the sculptor of the Crystal Palace dinosaurs in England.

What was the **first mounted dinosaur skeleton** in the Western Hemisphere?

In the Western Hemisphere, the first skeleton composed of real dinosaur bone fossils was mounted in 1901 at Yale's Peabody Museum of Natural History. A skeleton of an *Edmontosaurus* was mounted in an erect, bipedal, running stance.

What were the great North American **"Bone Wars"** of the late 1800s?

The "Bone Wars" was the name given to the great rush to find, collect, name, and describe dinosaur fossils discovered from areas in the western United States. The impetus to this fervor was the intense, bitter, personal rivalry between two American paleontologists: Othniel Charles Marsh (1831–1899) and Edward Drinker Cope (1840–1897).

Starting in the 1870s, both men—once friends—funded and led competing expeditions to sites in the western United States. Each was trying to discover and site was called the Carnegie Quarry). On October 4, 1915, President Woodrow Wilson designated the spot as Dinosaur National Monument—because of its importance to paleontology and to stop any future development of the area.

In 1909, Douglass found the dorsal bones of an *Apatosaurus* at this site; it took six more years to remove the skeleton from the rock and mount it at the Carnegie Museum. After 1922, Douglass worked the quarry for two more years (for the University of Utah and the Smithsonian Institution) finding a *Diplodocus* that is currently mounted at the Smithsonian. Today, tourists at the visitors center at Hogback Ridge can view the quarry face that acts as the north wall of the building; the bones were left in place after the overlying rocks were removed.

Where and what is **Como Bluff**?

Como Bluff is a long, east–west oriented ridge located in southern Wyoming. It is also the site of a famous Jurassic dinosaur fossil bed excavated during the great "Bone Wars" of the late 1800s. It was discovered by two employees of the Union Pacific Railway, W.E. Carlin and Bill Reed, as a new rail line was being built through the general area. They secretly contacted Othniel Charles Marsh (1831–1899), trying to sell him some gigantic bones. Marsh subsequently sent his assistant S.W. Williston (1851–1918) to investigate the situation. Williston informed Marsh that the bones "extend for seven miles and are by the ton.... The bones are very thick, well preserved, and easy to get out." Because of Williston's words, Marsh hired Carlin and Reed to work the beds exclusively for him and to send the fossil bones back to Yale University. From samples of bones uncovered at Como Bluff, Marsh named the dinosaurs *Stegosaurus, Allosaurus, Nanosaurus, Camptosaurus,* and *Brontosaurus* (now known to be the same animal as *Apatosaurus*). Excavations at Como Bluff were discontinued after 1889.

When was the **first complete fossil** skeleton of a ***Stegosaurus*** found in the United States?

In 1886, near Canyon City, Colorado, the fossilized remains of a *Stegosaurus* were found by Othniel Charles Marsh's crew. The animal's dorsal armor plates were arranged in two rows along the back, with the plates alternating position. This skeleton was subsequently displayed in the Smithsonian Institution in Washington, D.C., exactly as it was found in the field.

Where is the **largest known collection** of **theropod skeletons** in the world?

The largest mass accumulation of theropod skeletons in the world was found at the Ghost Ranch quarry in northwestern New Mexico. In 1947, George Whitaker and Edwin Harris Colbert (1905–2001), members of an expedition from the American Museum of Natural History, found over 100 skeletons of the Late Triassic period dinosaur *Coelophysis*. They found the dinosaur skeletons in Arroyo Seco ("Dry Canyon") on the lands of the Ghost Ranch.

There are a number of sites aroudn the world considered to be "fossil parks," places rich in large collections of dinosaur fossils (iStock).

The skulls in this bone bed show considerable variations in size, ranging from 3 to 10 inches (8 to 26 centimeters) in length, indicating the presence of both juveniles and adults. In 1948, during further excavations, George Whitaker and Carl Sorenson discovered two skeletons of *Coelophysis* with juveniles inside the stomach areas. This was thought to indicate that this species was cannibalistic.

It is still a mystery why so many of these animals ended up in such a small area. Some paleontologists suggest that a herd of *Coelophysis* was overwhelmed by a flood, perhaps while crossing a river. If this is true, then this discovery is the first evidence for herding behavior among a dinosaur species.

When was the **first *Triceratops*** fossil discovered in the United States?

A fossilized skull of a three-horned, Cretaceous period, herbivorous dinosaur was discovered by John Bell Hatcher (1861–1904) and Othniel Charles Marsh (1831–1899) in 1888 in the Judith River beds of Montana. This dinosaur would subsequently be named *Triceratops*.

When was the **first *Tyrannosaurus rex*** fossil discovered in the **United States**?

The first *Tyrannosaurus rex* skeleton was discovered in the United States in 1902 by Barnum Brown (1873–1963), who was perhaps the greatest collector of dinosaur

And for those who are unable to visit Chicago, replicas were made, which tour across the United States and internationally. Another life-sized cast of Sue is on display at DinoLand U.S.A. at Walt Disney World's Animal Kingdom in Orlando, Florida.

What are some other notable ***Tyrannosaurus rex* fossils** found in the **United States**?

The current size champion was found in the summer of 1997 in a Late Cretaceous period bone bed near the Fort Peck reservoir in Montana. This area is in the Badlands of eastern Montana. The remains were found in the Hell Creek rock layer, a geological formation well known for its dinosaur bones. The site appears to have been a former river channel. The bones of dead dinosaurs were washed into the channel and collected in one place. This skeleton, though only partially excavated, appears to be nearly complete and is the largest specimen of a *Tyrannosaurus rex* yet found. Its pubis bone is at least 52 inches (133 centimeters) long. The previous largest known *Tyrannosaurus rex* skeleton had a pubis bone approximately 48 inches (122 centimeters) long. The skull of this animal measures approximately 6.6 feet (2 meters) long.

There was another astounding discovery in 2001: a half-complete skeleton of a juvenile *Tyrannosaurus* was discovered in the Hell Creek formation in Montana, by a crew from the Burpee Museum of Natural History of Rockford, Illinois. It was named "Jane the Rockford T-Rex" and was initially considered the first known skeleton of the pygmy tyrannosaurid *Nanotyrannus*. After several leading experts examined the fossils, however, the consensus is that it is probably a juvenile *Tyrannosaurus rex,* the most complete and best-preserved to date.

There are also U.S. dinosaur fossils that still have to be confirmed and/or analyzed. For example, in 2001, Jack Horner revealed that he had discovered a *Tyrannosaurus rex* around 10 percent larger than the dinosaur "Sue" that is at the Field Museum in Chicago. Horner calls the specimen *C. rex* (or "Celeste," after his wife), but there is still work to do on the fossil.

Although discovered in the 1960s, researchers at Montana State University in 2006 claimed to have found the largest *Tyrannosaurus* skull yet. It measures 59 inches (150 centimeters) long compared to the 55.4 inches (141 centimeters) of "Sue's" skull, making it the largest discovered so far.

Which states in the United States have the **most dinosaur fossil sites**?

Although many states have dinosaur sites, so far the states that seem to be the most prolific producers of fossils are Colorado, Utah, Wyoming, and Montana.

Have **dinosaur fossils** been found in any U.S. **metropolitan areas**?

Dinosaur fossils have been found in the Denver, Colorado, metropolitan area that were exposed during recent construction projects. This urban area lies on top of a treasure trove of fossils, both dinosaur and others. Evidence suggests that millions of years ago this area was part of a tropical rain forest. During excavation for Denver International

What special tool was used to uncover the oldest dinosaur eggshell ever found in New Mexico?

The special tool used to uncover a one-square-inch (6.5-square-centimeter), 150-million-year-old dinosaur eggshell was a toy backhoe. This "sophisticated paleontological device" was wielded by three-year-old David Shiffler of New Mexico. His family had stopped along the Rio Puerco after a camping trip. Seizing the opportunity, David began to dig in a sandy spot with his toy backhoe, uncovering a small object he declared was a dinosaur eggshell. David's father later took the object to paleontologists, who confirmed that it was indeed a dinosaur eggshell from the Late Jurassic. This makes the dinosaur eggshell the oldest one found in New Mexico, and one of the oldest found anywhere. It also means that David is probably the youngest person ever to make a major paleontological discovery.

Airport in 1989, bone and plant fossils were found. There are other examples, too. Recently, the remains of four *Triceratops* were discovered during the construction of fairways at The Heritage at Westmoor Golf Course; and at a subdivision south of the golf course bones from five dinosaurs, a crocodile, and a mammal were found.

What **heavily armored dinosaur fossils** were found in **Utah**?

Scientists recently discovered two new species of heavily armored dinosaurs (or ankylosaurids) in Utah about 100 miles southwest of Salt Lake City. The species were both about 30 feet (9 meters) in length. One is an ankylosaur, or club-tailed armored dinosaur, and is the oldest ever found. The other fossil is that of a nodosaur, a clubless armored dinosaur, the largest on record. Most ankylosaurids were from Asia; these animals were thought to have crossed over a land bridge to North America about 100 million years ago.

Where was the world's **oldest** and most **primitive duck-billed dinosaur** found?

The world's oldest and most primitive duck-billed dinosaur, *Protohadros byrdi,* was discovered in a road cut near Flower Mound in northcentral Texas. When this dinosaur died almost 95.5 million years ago, the middle of North America had a shallow seaway, and northern Texas was a wooded marsh. Scientists are currently trying to determine if this means that the birthplace of hadrosaurs was North America instead of Asia, as previously thought.

Were **east coast dinosaur tracks** found by an **amateur paleontologist**?

Yes, a large number of dinosaur and flying reptile tracks from around 105 to 115 million years ago were found in local stream beds by Ray Stanford, an amateur pale-

Who was Barnum Brown?

Barnum Brown (1873–1963)—named after the famed circus showman P.T. Barnum—was perhaps the greatest collector of dinosaur fossils ever known. Based out of the American Museum of Natural History, Brown traversed the country, bargaining and trading for fossils of both dinosaur and other species. One of his best dinosaur discoveries was made in 1910: several hind feet from a group of *Albertosaurus* collected in Dry Island Buffalo Jump Provincial Park in central Alberta, Canada.

land and in water), but rather a land-dwelling animal with habits similar to those of the modern elephant.

What **discovery by John Ostrom** led to new theories about dinosaur **behavior and physiology**?

The discovery and description of *Deinonychus,* "terrible claw," by John Ostrom (1928–2005) of the Yale Peabody Museum turned out to be the catalyst that changed our perceptions about dinosaurs. In 1964, *Deinonychus* bones were excavated from the Cloverly formation rocks of the Early Cretaceous period in Montana; Ostrom presented his findings in 1969. The information from these fossils, and other fossil finds related to *Deinonychus,* increased our knowledge of dromaeosaurids, which may have been the most aggressive and maybe the most intelligent of the theropods.

Based on fossil evidence found during the excavation, Ostrom concluded that these animals may have hunted in packs, indicating a social structure. Also, the animals' skeletons were light and slender, with a stiffened tail for balance, long clawed arms for grasping, sharp backward-curving teeth for tearing flesh, and a huge sickle-shaped claw on the second toe of the foot for slashing. The animal was built for speed and agility, quite unlike the perception of dinosaurs up until that time. From these findings, Ostrom theorized that *Deinonychus* may have been warm-blooded. This radical notion invited new thinking about dinosaur physiology and led to the modern ideas of dinosaurs as active, social animals.

But Ostrom's ideas did not stop there; he was also the paleontologist who almost single-handedly convinced the scientific community that birds are descended from dinosaurs. In addition, his discoveries provided the underpinning for the *Jurassic Park* books and movies, as well as tons of books about dinosaur evolution.

Who is **John R. Horner**?

In 1978, American paleontologists John R. Horner (1946–) and Bob Makela discovered the fossilized remains of what would subsequently be called *Maiasaura,* or "good mother lizard," in Montana. This was the first known nest of baby dinosaurs, and indicated the young had been cared for by adult dinosaurs. Starting in 1979, and working into the 1980s, Horner uncovered evidence of herding behavior in the dinosaurs, as well as nesting grounds, providing new insights into the social behavior of these dinosaurs. The herd was estimated to have been almost 10,000 dinosaurs strong.

Currently, John R. Horner is the regents' professor and curator of paleontology for the Museum of the Rockies at Montana State University in Bozeman, Montana. He is one of the most famous North American paleontologists, known for his dinosaur fossil discoveries and for being the consultant for Hollywood's *Jurassic Park* films.

Who is **Robert Bakker**?

Robert Bakker (1945–) is an American paleontologist; he was previously at the Morrison Natural History Museum in Morrison, Colorado, but more recently lectures and hunts for dinosaur fossils at Como Bluff, Wyoming. Like his mentor John Ostrom, Bakker has helped reshape modern theories about dinosaurs, particularly arguing in favor of the theory that some dinosaurs were warm-blooded (endothermic); he has expanded and/or changed many conventional ideas about dinosaur behavior; and he was also one of the first paleontologists to suggest that many dinosaurs had feathers. He is the author of many books, including *Dinosaur Heresies, Raptor Red, Maximum Triceratops,* and *Raptor Pack*.

Who is **Jim Kirkland**?

James Kirkland (1954–) is an American paleontologist and geologist who has worked extensively with dinosaur fossils from the southwestern United States. He is responsible for discovering new and important genera; for example, in 1991, Kirkland found the first skeleton of *Utahraptor,* a large dromaeosaurid with long foot claws, in the Gaston Quarry, Utah. He currently is an adjunct professor of geology at Mesa State College in Grand Junction, Colorado. He is also a research associate at the Denver Museum of Natural History in the Denver Museum of Nature and Science, as well as official Utah State Paleontologist for the Utah Geological Survey.

Who is **Paul C. Sereno**?

Paul C. Sereno (1958–) is an American paleontologist currently at the University of Chicago. He has worked at many dinosaur dig sites, including those in South America, Asia, and Africa. Among his dinosaur accomplishments, he: discovered the first complete skull of *Herrerasaurus;* was responsible for excavating a giant *Carcharodontosaurus* in 1996; named the oldest-known dinosaur, the *Eoraptor* (with others) in 1993; and, in 1991, found the second oldest fossil birds, *Sinornis* ("Chinese bird"). He also named many new dinosaur fossils and rearranged the dinosaur family tree, reorganizing the ornithischians and naming the clade Cerapoda (formed from the ornithopods and marginocephalians).

What is the *Scrotum humanum* and when was it discovered?

Around 1676, the Reverend Plot of England reported a fossil bone find: a "human thigh bone of one of the giants mentioned in the Bible." In 1763, R. Brookes named the bone based on its shape. He referred to it as *Scrotum humanum*—the genitals of a giant man. Today, the bone is thought to be from a megalosaurid dinosaur, and is from the distal end of a femur. The name *Scrotum humanum*—though unofficially the first assigned name given to dinosaur bones—has never been used, since it was an erroneous designation.

lished his studies of a Cretaceous period carnivore whose fossils had been found in Stonesfield, England, describing the animal based on fossilized jaws and teeth. He also used the name *Megalosaurus,* presenting his data at a meeting of the Geological Society of London. This was subsequently accepted as the first dinosaur to be described.

Who **first coined** the word **palaeontology**?

In 1830, Sir Charles Lyell (1797–1875), a Scottish geologist, coined the word palaeontology, or "discourse on ancient things." Between 1829 and 1833, Lyell recognized that this was a separate field of science. In general, "paleontology" is the spelling used in the United States for this field of study.

Who was the first person to **coin** the term **"dinosaur"**?

The first person to realize the bones of the ancient giant reptiles belonged to their own unique group was Sir Richard Owen (1804–1892), an English anatomist and paleontologist. In 1841, based on several partial fossil remains of *Iguanodon, Megalosaurus,* and *Hylaeosaurus,* he coined the term "dinosaur" ("terrible lizard") to describe them.

Where were the **first life-sized dinosaur models** publicly displayed?

The first life-sized dinosaur models were publicly displayed at Sydenham Park, site of the relocated Crystal Palace in southeast London, England. The year was 1854, and Sir Richard Owen supervised the sculpting of these figures by Benjamin Waterhouse Hawkins (1807–1889). The figures were placed in the park, later renamed the Crystal Palace Park; the dinosaurs were all portrayed as giant elephantine lizards. These figures were enormously popular with the public. Although the Crystal Palace itself burned, the sculptured dinosaurs survived the fire and can still be viewed on the grounds.

Who **first proposed extinctions** had occurred during Earth's history?

Baron Georges Cuvier (1769–1832), a French scientist at the National Museum of Natural History in Paris, was the first to propose the idea of extinctions. Around

1800, his work with mammoth and mastodon bones (although they were not labeled as such until later) found in North America led to his theory of extinction. He showed that these creatures recently went extinct, refuting claims that all creatures that ever existed were still alive today and that fossils were just evidence of as-yet-undiscovered species living elsewhere on the planet. Cuvier is considered to be the father of modern paleontology and comparative anatomy. His confirmation of the extinction process led to the study of even more ancient animals—the dinosaurs.

French scientist Georges Cuvier came up with the theory that species go extinct, which was counter to the commonly accepted idea that the species we see today have always been around (iStock).

Who was first to reconstruct the way **dinosaurs behaved**?

Louis Dollo (1857–1931), a French mining engineer, was the first to interpret the remains of dinosaurs with an eye toward reconstructing their lifestyles. In 1878, the remains of approximately 40 *Iguanodon* were discovered in the Fosse Sainte-Barbe coal mine near the town Bernissart in southwestern Belgium. Their excavation took three years. Dollo spent the rest of his life assembling, studying, and interpreting the fossil remains. In 1882, he started as an assistant naturalist at the Royal Natural History Museum in Brussels, Belgium; in 1904, he became director of the museum, based on the strength of his scientific discoveries associated with the *Iguanodon*. Not only did he assemble the bones for exhibit and write papers concerning his findings, but he also attempted to describe the behavior of these animals.

Who **first classified dinosaurs** based on the structure of the **pelvic area**?

In 1887, English paleontologist Harry Seeley (1839–1909) realized there were two distinct groups of dinosaurs. He classified them as *Ornithischia* (bird-hipped) and *Saurischia* (lizard-hipped), basing this primarily on the bone structure of the pelvic area. This system of classification was widely adopted and is still in use today.

Why are the **Solnhofen quarries** of Germany so **important** to paleontology?

The Solnhofen quarries of Bavaria, Germany, are important not just because they are home to the oldest known bird fossils, the *Archaeopteryx lithographica*. This

Famed paleontologist Richard Owen designed these dinosaur statues in 1852; they are still on display at London's Crystal Palace Park (Big Stock Photo).

area is what paleontologists call a *Lagerstatten* ("fossil lode" or "storehouse"). Because of their unique prehistoric conditions, these sites have preserved numerous animals, giving us a virtual snap-shot of the fauna during this time. There are only approximately 100 fossil sites around the world designated as *Lagerstatten,* with each representing different time periods, and all rich in fossil varieties.

The quarries were the site of a quiet, warm-water, anoxic (lacking oxygen) lagoon. It lay behind reefs on the northern shores of the Tethys Sea approximately 150 million years ago. The tropical climate at this time was perfect for the animals and plants living along its shores. And the ocean itself teemed with life beyond the stagnant lagoon. Storms would sweep in dead or dying animals from the ocean; dying land creatures either fell into the lagoon, or drifted into it from the shore. Their bodies fell to the bottom of the lagoon and were covered by soft lime mud; little oxygen was present to decompose the organisms. The ensuing fine limestone rock preserved, in exquisite detail, the remains of over 600 species, including the smallest dinosaur, *Compsognathus,* pterosaurs by the hundreds, numerous insects, and, of course, the remains of *Archaeopteryx lithographica*.

When was the **most spectacular** dinosaur fossil **expedition** mounted?

The most spectacular dinosaur fossil expedition ever mounted began in 1909 and lasted through 1912. The expedition took place in German East Africa (now Tanzania) around the village of Tendaguru.

In 1907, W.B. Sattler found gigantic fossil bones weathering out of the surface rock as he explored the area around Tendaguru for mineral resources. After Sattler reported his findings, a noted paleontologist, Professor Eberhard Fraas (1862–1915), visited the area and took collected samples back to Germany. There, Dr. W. Branca, the director of the Berlin Museum, realized the importance and scope of the findings and started raising funds for an expedition.

The expedition began in 1909; it was a search larger in scope than anything to date. In the first year, 170 native laborers were employed; in the second year, 400 were used. The third and fourth years saw 500 natives at work on the dig sites, which were located in an area extending almost two miles (three kilometers) between Tendaguru Village and Tendaguru Hill. The laborers were accompanied by their families; thus, the expedition had to accommodate upwards of 700 to 900 people. Additionally, after the fossils were mapped, measured, excavated, and encased in plaster, they had to be hand carried from Tendaguru, in the interior, to Lindi, on the coast—a trek that took four days. There, the enormous number of bones, eventually totaling 250 tons (230 metric tons), were shipped to Germany for preparation, study, and reconstruction.

What dinosaur remains were excavated during the **Tendaguru expedition**?

The Tendaguru expedition was itself spectacular, and so were the dinosaur fossils discovered at the site. Among the findings were three types of theropods—the small, agile *Elaphrosaurus,* and the larger *Ceratosaurus* and *Allosaurus*. Six herbivorous dinosaurs were also found: the tiny ornithopod *Dryosaurus; Kentrosaurus,* a stegosaur; and four sauropods, *Dicraeosaurus, Barosaurus, Tornieria,* and the largest one of the time, *Brachiosaurus*. The reconstructed skeleton of a *Brachiosaurus* in the Berlin Museum from this site is the largest complete dinosaur skeleton in the world.

In addition to these spectacular dinosaur finds, the expedition also uncovered remains of pterosaurs, fish, and a tiny mammal jaw bone. All of the animals were similar to those found earlier in the Morrison formation in the western United States, indicating that migration between North America and Africa was relatively easy during this time.

Where were the **first *Albertosaurus*** fossil **remains** found?

The first fossil remains of the Late Cretaceous dinosaur *Albertosaurus* were found in the Badlands of the Red Deer River Valley of Alberta, Canada. Geologist Joseph Burr Tyrrell found the fossil remains in the spring of 1884, as he led an expedition near present day Drumheller for the Geological Survey of Canada.

In the early 1900s, the discovery of these and other remains brought numerous paleontologists to the area, including Barnum Brown of the American Museum of Natural History, and Charles H. Sternberg and sons for the Canadian Geological Survey. The friendly rivalry between Brown and the Sternbergs was dubbed the "Great Canadian Dinosaur Rush."

The first fossils of the Cretaceous dinosaur *Albertosaurus* were found in Alberta, Canada, which is how this predator got its name (Big Stock Photo).

Today, the Badlands of the Red Deer River Valley in Alberta, Canada, are recognized as one of the world's leading fossil collecting areas, with some 25 species of dinosaurs so far uncovered. The significance of this area led to the establishment of the Royal Tyrrell Museum, established in June 1990, in Drumheller.

How did the **"Badlands"** of the Red Deer River Valley **form**?

The badlands of the Red Deer River Valley, in Alberta, Canada, were carved by meltwater torrents when the ice sheets retreated approximately 10,000 to 15,000 years ago. There is some evidence that flash floods, rather than rivers, were the agents that created the present Badlands topography. These landscapes include narrow, winding gullies and channels; heavy erosion; steep slopes; and little or no vegetation.

During the time of the dinosaurs, this area included numerous deltas and river flood plains that extended out into a shallow, inland sea. The Late Cretaceous deposits of sand and mud often included the bodies of dinosaurs. Over millions of years, as material was laid down layer upon layer, the deposits turned into rock, fossilizing the dinosaur bones.

The advance and retreat of four glacial ice sheets over millions of years—along with other natural erosion processes by wind and water—caused significant wearing away of the area. The material on top was removed, and the exposed Cretaceous period sedimentary rocks were carved into the Badlands of today. The Cretaceous layer is known as the Horseshoe Canyon formation and is continually eroding, exposing fresh dinosaur fossils.

DINOSAUR DISCOVERIES

What great **Jurassic** and early **Cretaceous** period **discovery** was made in **Asia**?

The site of spectacular fossils, including many dinosaur remains, was found in Liaoning Province, northeast China, near the village of Beipiaon. The site contained the first fossilized internal organs of dinosaurs and the first fossil of a dinosaur containing the remains of a mammal it might have eaten. This site has also yielded the remains of *Confuciusornis,* one of the oldest beaked bird; the feathered dinosaur *Sinosauropteryx prima;* the oldest modern bird, *Liaoningornis;* a dinosaur found in a sleeping position; a *Protarchaeopteryx,* which is a primitive bird perhaps older than *Archaeopteryx lithographica;* the earliest flower, placental mammal, and marsupial; and many other species of dinosaurs, mammals, insects, and plants. There are new findings every year.

The fossils found at the site were preserved in great detail because the prolific rock layer is from a lacustrine (lake) deposit that was covered with a fine volcanic ash. Paleontologists speculate that a brief catastrophe, such as a volcanic eruption, killed and quickly buried everything in the area. Thus, even impressions of soft body parts, such as feathers and organs, were preserved.

Where do **Cretaceous dinosaur fossils** continue to be found in **Mongolia**?

At Ukhaa Tolgod, in the Gobi Desert of Mongolia, lies what is billed as one of the greatest Cretaceous fossil finds in history. Starting in 1993, the discoveries have included the remains of more than 13 troodontid skeletons, over 100 uncollected dinosaur specimens; numerous mammals, and a nest-brooding adult *Oviraptor*. The reason for the huge number and extraordinary states of preservation is thought to be due to a series of catastrophic occurrences. These events swiftly buried the animals, precluding any damage by the elements or scavengers. Scientists believe normally stable sand dunes became drenched with rain water, triggering sudden debris flows that trapped—and preserved—the animals.

What is one of the **most famous dinosaur sites** in the **world** today?

Flaming Cliffs, located in the Gobi Desert of Mongolia, is one of the most famous dinosaur sites in the world today. In the 1920s, an expedition from the American Museum of Natural History, led by Roy Chapman Andrews, yielded the first dinosaur eggs at this site. It was closed for many years, but in the late 1980s the site was reopened to scientific study and dinosaur fossil gathering. Since that time, numerous findings have continued to occur in this dinosaur-rich area.

While many dinosaur fossils have been uncovered in the western United States, important finds are being made in such countries as China and, shown here, remote Mongolia. Government restrictions can make obtaining permits for digs in these areas very problematic (iStock).

What **Chinese dinosaur fossils** may be the closest-known **ancestor of birds** yet found?

The *Caudipteryx* ("tail feather") was a 3-foot- (1-meter-) tall, feathered theropod dinosaur that may be the closest-known ancestor of birds to date. Fossils of this creature have been uncovered in the sediment of an ancient lake bed in China's Liaoning Province. This dinosaur lived during the Late Jurassic to Early Cretaceous periods (about 120 to 136 million years ago) and had asymmetrical feathers, indicating that it did not fly, but used the feathers for insulation. Unlike other theropod dinosaurs, the *Caudipteryx* did not have a long tail, but it did support feathers up to 8 inches (20 centimeters) in length.

What do certain **fossils in China** indicate about **dinosaur social groups**?

The fossilized remains of six infant dinosaurs that died in a volcanic mudflow were recently found in China. Researchers say the animals were less than four years old and probably formed a "crèche" composed of babies from at least two different clutches. The discovery of these 120-million-year-old *Psittacosaurus* may mean that the animals formed social groups much earlier than previously thought.

What events in **Russia** have highlighted **crime** in dinosaur **fossil collecting**?

In 1996, the remains of five dinosaurs disappeared from the fossil repository of the Paleontological Institute of the Russian Academy of Sciences, Moscow, and are

On the coast of bucolic Gippsland, Australia, the first evidence of dinosaurs on that continent was discovered. This discovery was of particular interest to scientists because at the time the dinosaurs were alive Australia was a frozen wilderness joined to Antarctica (iStock).

believed to have been stolen. They included a lower jaw and maxilla with teeth from the large carnivore *Tarbosaurus efremovi;* a skull of *Breviceratops kozlowskii,* a late Cretaceous herbivore; and two skulls of *Protoceratops*.

Who found the **first evidence of dinosaurs** in **Australia**?

The first evidence of dinosaurs in Australia was found in 1906 by a Mr. Ferguson, a government geologist: A small carnivorous dinosaur claw was discovered on the Gippsland coast of eastern Victoria. Since then, many more dinosaurs have been discovered in places such as Queensland, South Australia, and New South Wales. These discoveries include sauropods and iguanadons, such as the *Muttaburrasaurus,* which is unique to Australia.

Why is it interesting that **dinosaurs** lived in **Australia**?

One reason that it is interesting that dinosaurs lived in Australia is the climate. Australia, during the age of the dinosaur, was very cold. This is because millions of years ago, Australia was within the polar circle and still attached to Antarctica.

What **dinosaur fossils** were **recently found** in **Queensland, Australia**?

Three special dinosaur fossils—one theropod and two long-necked sauropods—were announced in 2009, the biggest dinosaur discoveries in Australia in decades. Nicknamed Banjo, Matilda, and Clancy, the fossils were uncovered near a small lake, or billabong, in Queensland. It's estimated that the dinosaurs lived about 98 million years ago. Scientists believe the carnivorous dinosaur Banjo (*Australovenator win-*

In what northern locale were dinosaur footprints found in 1960?

The Arctic island of West Spitzbergen (Svalbard) is where an international team of geologists discovered dinosaur footprints in 1960. These tracks were thought to have been made by an *Iguanodon,* an early Cretaceous period dinosaur.

tonensis) probably rivaled the ferocious, well known *Velociraptor,* but was much larger. Plant-eaters Matilda and Clancy were titanosaurs, which are thought to be some of the heaviest dinosaurs that roamed the Earth.

What fossil was **stolen** from a site in **Australia**?

In 1996, fossil collectors stole a dinosaur footprint from a site in northwest Australia, approximately 1,800 miles (3,000 kilometers) from Sydney. This was the world's only known fossil footprint made by a *Stegosaurus,* and represented a loss not only to science, but to the aborigines of Australia, who regarded the location of the footprints as one of their sacred sites. The thieves apparently used power tools to remove the rock containing the trace fossil. Under aboriginal law, this offense is punishable by death.

Approximately one year after thieves made off with the fossil, police recovered it and arrested two Australian men. The recovered block of rock containing the footprint weighed approximately 66 pounds (30 kilograms), measured 23 inches (60 centimeters) by 15.5 inches (40 centimeters), and was 5 inches (13 centimeters) deep. The men attempted to sell the fossil footprint in Asia, but they were unsuccessful. Police believe this may have been due to the size or weight of the rock. Police also would not elaborate on how the fossil was recovered.

Where were the **first duck-billed dinosaur fossils** found outside the Americas?

The first duck-billed dinosaur (plant-eating hadrosaurs) fossils found outside the Americas were discovered on Vega Island, which is off the eastern side of the Antarctic Peninsula. A tooth found there was dated at approximately 66 to 67 million years old. This finding gives more credence to theories about a land bridge connecting South America and Antarctica during this time. It also indicates that this cold climate and ecosystem were once lush and robust enough to support large plant eaters.

What **dinosaurs** have been discovered in **Antarctica**?

Along with the hadrosaur mentioned above, there have been several discoveries of dinosaurs in Antarctica. For example, a 12-foot- (4-meter-) long, bipedal, plant-eating iguanodontid was discovered in February 1999 on a rocky beach of James Ross

Island. A 70-million-year-old bipedal, meat-eating dinosaur (an unnamed theropod) was also discovered on James Ross Island in late 2003. The discoveries included a lower leg and foot bones, fragments of the upper jaw, and some teeth. Still another dinosaur, a 200-million-year-old bipedal, long-necked, plant-eating, unnamed sauropod was discovered in 2003 in the Antarctic interior, including fossils of a pelvis and one of the dinosaur's hipbones.

What **English island** continues to be a hotspot for **dinosaur fossil discoveries**?

The small Isle of Wight, 3 miles (4.8 kilometers) off the southern coast of England, became a major hotspot for paleontologists, and it continues to be one of the world's best dinosaur fossil discovery sites. This island was once home to Queen Victoria, and is best known as a holiday destination. But the quality of its dinosaur fossils—and the time period from which they originate—make this island the focus of paleontologists worldwide. The uniqueness of the dinosaur fossils found on the Isle of Wight has to do with their numbers and age: over 20 species of dinosaur having been recognized from the Early Cretaceous (in particular between 110 and 132 million years ago). Fossils from this period are rare; most sites around the world yield fossils from the Triassic or Jurassic periods. Thus, the Isle of Wight dinosaur fossils are essentially a window into a unique period of time, and the island constitutes a resource not found elsewhere in the world. In addition, the fossils are well-preserved and articulated, meaning the bones are still joined together, not strewn around.

Why are **dinosaur fossils** relatively easy to find on the **Isle of Wight**?

The dinosaur fossils on the Isle of Wight are relatively easy to find because they are concentrated in two small areas. There is a 6-mile- (10-kilometer-) long fossil-containing area along the southern coast of the island, and another half-mile- (0.8-kilometer-) long area on the eastern coast. In many other of the top fossil sites in the world, the bones of dinosaurs are distributed over thousands of square miles, making the search for them more diffuse and difficult.

The continual uncovering of dinosaur bones on the Isle of Wight is due to a unique combination of physical factors. The rock containing the dinosaur bones has been pushed to the surface by geological pressures; in other parts of the world, this type of rock may be buried underground—sometimes by as much as 1,000 feet (305 meters) below the surface.

The fossil-bearing rock itself is a soft rock, consisting mainly of sandstones and mudstones, which makes the rock susceptible to erosion. And the Isle of Wight experiences extensive erosion. More than three feet (one meter) of rock is worn away every year in certain exposed areas. The engine for this high rate of erosion is the sea. Every fall and winter, the flow of the tides, combined with high seas and gales from the English Channel, batter the rocks that make up the sea cliffs on the island. But this has been a boon to scientists: the erosion exposes dinosaur bones, which literally fall onto the beaches at the feet of eagerly awaiting paleontologists.

Two factors make dinosaur fossil hunting on England's Isle of Wight rather easy: the fossils are concentrated in two small areas, and the island is made up of soft rock that is easy to excavate, such as the chalk on these cliff faces (iStock).

Among the many exciting finds on the Isle of Wight are: the skeleton of the first small, meat-eating dinosaur found in England, a new species, was found in a crumbling cliff by an amateur collector; in 1997, a smaller version of a *Tyrannosaurus rex,* another new species called *Neovenator salerii,* was also discovered; and a very well preserved *Iguanodon* was uncovered. The fossil remains of a 12-foot (4-meter) carnivorous dinosaur were discovered—a cat-like creature, with unusually long hind legs that permitted it to run at high speeds; it was also equipped with claws and razor-sharp teeth and became the first small meat-eating dinosaur found in Britain.

What was *Valdosaurus*?

Valdosaurus ("Wealden Lizard") was a small- to medium-sized ornithopod dinosaur whose remains have been found on the Isle of Wight. Only a few partial, fragmented remains have been found over the years. What little evidence there is suggests that this dinosaur was about 14 feet (4.25 meters) long and about 4 feet (1.2 meters) high at the hips. The excavated leg bones of *Valdosaurus* are comparable to those of *Dryosaurus,* a dryosaurid (oak lizard) known from the Late Jurassic period of North America, which lived some 30 million years earlier. Using *Dryosaurus* as a guide, *Valdosaurus* may have been a fast-running biped with long hind legs, short front limbs, a small skull, and stocky build. From the shape and size of the hind limb bones, we know that *Valdosaurus* was an agile and fast runner; there is currently no evidence of any armor plates, sharp teeth, horns, or

Where has a two-ton dinosaur pelvis been discovered?

A two-ton dinosaur pelvis dating from the upper Jurassic period has been discovered at a dig site near Lourinha, Portugal, located about 37 miles (60 kilometers) north of Lisbon. Evidence of dinosaurs is prevalent in this area of central Portugal, with many recent discoveries of tracks and skeletons. Dinosaur eggs with fossilized embryos have also been found here. Paleontologists believe the pelvis came from a sauropod about 66 feet (20 meters) tall and weighing 20 tons (18 metric tons). This pelvic bone discovery will help shed light on the posture of dinosaurs.

claws. Thus *Valdosaurus* probably was a grazing dinosaur, eating ferns and cycad-like fronds. But it remained continually alert for predators, using its running ability as a defense against attacks.

Where were the **oldest fossilized dinosaur embryos** found?

The oldest fossil embryos were discovered among approximately 100 dinosaur eggs recently found in the area around Lourinha, a small town about 37 miles (60 kilometers) north of Lisbon, Portugal. The embryos—identified as those of theropods—were dated at approximately 140 million years old. They are the oldest embryos to date, and the only ones currently known from the Jurassic period. Until this finding, all other fossilized dinosaur embryos came from the Cretaceous period, with the oldest being approximately 80 million years old.

Where is one of the **world's largest paleontological institutes** located?

One of the world's largest paleontological institutes is located in Moscow, Russia, and is known as the Paleontological Institute of the Russian Academy of Science. No other institute in the world has more paleontologists under one roof, with research interests ranging from the dinosaurs of Mongolia and mammals from Georgia (near Russia) to the origins of life itself. There are extensive collections of fossils from all over the former Soviet Union and the world, along with a Museum of Paleontology and great public exhibits. The exhibits include Mongolian dinosaurs, synapsids from the Perm region, and Precambrian fossils from Siberia. Unfortunately, with all of these wonderful resources and exhibits, the Institute and Museum are largely underfunded and remain relatively unknown to those outside the paleontological world, unlike more popular museums such as the American Museum of Natural History.

Where were the remains of the **oldest-known dinosaur** discovered?

The fossil remains of *Eoraptor* ("dawn hunter") were discovered in the Ischigualasto formation rock layer in Argentina, South America, which is the same area where the second oldest dinosaur, the *Herrerasaurus,* was found. In 1991, Ricardo Mar-

The main hall at the Paleontology Museum in Moscow boasts a full skeleton of a diplodocus. The Paleontological Institute of the Russian Academy of Science is the largest of its kind in the world (Big Stock Photo).

tinez made the first find of an *Eoraptor;* this was followed by the discovery of another skeleton in the 1990s by Fernando Novas and Paul Sereno. In 1993, the analysis of these remains led Sereno to name *Eoraptor* as the "first," or most primitive known, form of dinosaur. The *Eoraptor* was smaller than *Herrerasaurus,* being about three feet (one meter) long. It had all the dinosaur characteristics found in *Herrerasaurus,* but its skull was a basic design. It also had a few specializations that would allow it to be placed in any of the major dinosaur groups. The primitive dinosaurs probably represented five percent of the total animal population in the beginning of the Late Triassic period. But they would soon spread throughout the world and dominate the land during the Jurassic and Cretaceous periods.

Where was the **second oldest dinosaur** found?

In 1959, the remains of a *Herrerasaurus*—thought by some paleontologists to be the second oldest known dinosaur—was discovered in Argentina. It was found in a layer of rock known as the Ischigualasto formation; the area is located in the Ischigualasto Valley, or the "Valley of the Moon." The discoverers were a goat herder named Victorino Herrera and paleontologist Osvaldo Reig. In 1988, a complete skull and skeleton of *Herrerasaurus* was found in the same area by Paul Sereno and Fernando Novas. This 10- to 20-foot (3- to 6-meter) reptile had numerous characteristics enabling it to succeed at its carnivorous lifestyle, such as recurved teeth, powerful hind limbs, a bipedal stance, and strong arms. It was excavated from rock laid down during the beginning of the Late Triassic period, approximately 230 million years ago.

Where was a ***Giganotosaurus carolinii*** found?

A *Giganotosaurus carolinii,* one of the contenders for the world's largest carnivorous dinosaur, was discovered in Argentina by Ruben Carolini, an auto mechanic and amateur fossil hunter. Although this dinosaur had a similar appearance to *Tyrannosaurus rex*—it was slightly larger than *Tyrannosaurus rex* but had a much smaller brain (the size and shape of a banana)—paleontologists don't think the two were related.

The skeletal remains of a group of *Giganotosaurus* were discovered in Neuquen, a southern province of Argentina. There were bones of four or five of these Cretaceous period theropods; two were very large and the others smaller. Paleontologists believe this discovery is the most important evidence to date of a social, pack-hunting behavior on the part of large carnivorous dinosaurs. Approximately 90 million years ago, when these animals died, this region was similar to the present-day pampas of northern Argentina. The climate was warm and rainy, and the land was essentially low scrub, dotted with araucaria trees. After the group of *Giganotosaurus* died, a west-flowing river swept their bodies away. A local goatherder found the deposit containing the dinosaur bones in an area that is now a sandy rise in a desert.

Why was the discovery of ***Giganotosaurus*** **important**?

The discovery of *Giganotosaurus* was important because it cleared up a mystery: why *Tyrannosaurus rex* had never migrated into South America. Both of these dinosaurs occupied the same position on top of the food chain, and it now appears that the roughly equal-sized *Giganotosaurus* kept the *T. rex* out of South America. In turn, *T. rex* kept *Giganotosaurus* out of North America. Other dinosaurs, such as the smaller theropods and larger herbivores, freely wandered between the two landmasses where there was a land bridge between them.

What major **dinosaur egg** discovery was made in **South America**?

Thousands of fossilized dinosaur eggs—along with parts of teeth, skin, and bones from the unhatched embryos—were discovered in the northwestern Patagonian province of Neuquen, Argentina, in a place called Auca Mahuida. Paleontologists have nicknamed the site "Auca Mahuevo" after *huevo,* the Spanish word for "egg." Among the thousands of eggs are the first embryos ever found of a sauropod dinosaur, the large, four-footed plant-eaters. In addition, some eggs contained the first embryonic dinosaur skin ever found. Approximately 70 to 90 million years ago, this area of South America looked like the plains of the America Midwest. Now it resembles the Badlands of South Dakota, with erosion continually exposing rocks, bones, and eggs.

Paleontologists have used two clues to identify the type of dinosaur that laid the eggs at the recently discovered site in northwest Patagonia, Argentina: the embryonic teeth and the skin found in association with the eggs. The embryos had tiny, peg-like teeth, a characteristic of the sauropods. Paleontologists noted spots on the teeth had been rubbed flat by friction, indicating that the embryonic dinosaurs

What discovery shows a relationship between dinosaurs in the Northern and Southern Hemispheres?

The discovery of the foot of *Araucanoraptor argentynus* in Neuquen, Argentina, is another piece of evidence that dinosaurs migrated into the Southern Hemisphere and were, therefore, related to those found in the Northern Hemisphere. This particular dinosaur was a 90 million-year-old theropod originally thought to have never made it to the Southern Hemisphere. For a long time, paleontologists believed that dinosaurs from the Northern Hemisphere were completely different than those found in the Southern Hemisphere. Fossil finds like those of *Araucanoraptor argentynus* show that they were indeed related. In fact, *Araucanoraptor argentynus* was related to the *Troodon,* whose fossil remains have been found in Canada.

ground their teeth even before they were hatched. Some scientists believe this shows that the young were exercising their jaw muscles.

The embryonic skin found in some eggs has clearly visible scales. And based on the patterns in the skin, paleontologists believe the eggs were laid by titanosaurs. This species of sauropod dinosaur grew to approximately 45 feet (14 meters) long and were the only sauropods to survive to the end of the Cretaceous period.

Where were fossil remains of a ***Suchomimus tenerensis*** **found**?

The remains of a *Suchomimus tenerensis*—a large, fish-eating dinosaur with a long, crocodile-like snout—were found in the Tenere Desert of Niger, a central African country. Fossils of this Cretaceous dinosaur were located in rock strata called the Elrhaz formation. Approximately 100 million years ago, this area had a lush climate with plenty of water and supported many types of animals, including giant crocodiles, large fish, and many different species of dinosaurs such as the newly discovered *Suchomimus tenerensis*. Over time, the remains of this African dinosaur were swept into a river and covered by sediment. The winds of the desert eroded the sands covering the remains, exposing it to the light of day.

What **mysteries** did the remains of the dinosaur ***Majungatholus*** **solve**?

The discovery of *Majungatholus* on the island of Madagascar off the southeast coast of Africa cleared up three mysteries that had been puzzling paleontologists for years: 1) What dinosaur left behind numerous fossil teeth on Madagascar? 2) Why were there remains of a Northern Hemisphere pachycephalosaur on the island? 3) How did dinosaurs get from South America to Madagascar?

Paleontologists discovered hundreds of dagger-like fossil teeth throughout Madagascar over a hundred years ago, but no one knew what type of dinosaur had shed the teeth (during a meal, some carnivorous dinosaurs shed a few of their teeth in a manner similar to modern sharks and crocodiles). In 1996, an expedition went to Madagascar to find the dinosaur associated with these teeth. One day, while digging into a hill, a paleontologist found some tail bones. Further digging exposed an upper jaw bone of a large carnivorous dinosaur; the jaw contained the same teeth found scattered throughout Madagascar. The mystery had been solved.

The dinosaur responsible for the fossil teeth was named *Majungatholus,* a distant relative of *Tyrannosaurus rex*. It was approximately 20 to 30 feet (6 to 9 meters) long and lived in Madagascar about 70 million years ago. The skull of the dinosaur had a stubby remnant of a horn set between the eyes; some of the skull bones had an unusually rough texture, perhaps echoing patterns in the overlying skin. Paleontologists speculate that the combination of horn and texture on the head may have been used to threaten enemies or attract a mate.

How did the ***Majungatholus*** remains **explain** the **pachycephalosaur**?

At the turn of the century, fragments of a dinosaur skull were found in Madagascar. One of these fragments had a protrusion, which led some paleontologists to believe these were the remains of a pachycephalosaur, or "dome-headed" dinosaur. This group of dinosaurs was herbivorous and may have used their thickened skulls as battering rams. Scientists named this dinosaur *Majungatholus,* but for many years it was only known from a few fossil fragments. The problem was that pachycepthalosaurs have only been found in the Northern Hemisphere. How this dinosaur got to Madagascar was unknown.

With the recent discovery of more complete remains of a *Majungatholus,* the mystery was finally solved. The older skull fragments matched those of the newly discovered dinosaur. Paleontologists realized that the protrusion on the original bones had been wrongly identified as a dome when in reality it was a horn. There had not been a pachycepthalosaur in Madagascar after all. The *Majungatholus* was a large, carnivorous dinosaur with a small horn between its eyes.

How did the ***Majungatholus*** remains explain **dinosaur migration** from South America to Madagascar?

The *Majungatholus* was very similar to another dinosaur found in Argentina, although this animal had two horns. Other bone fragments found in India were also very similar, indicating these dinosaurs were all from the same group. However, no evidence has been found to date of any of these dinosaurs in Africa.

Approximately 120 million years ago, South America, Africa, Antarctica, Madagascar, Australia, and India were all joined together in one supercontinent called Gondwana, or Gondwanaland. Scientists originally believed a piece of the landmass containing South America and Africa first split away from Gondwanaland, then pulled apart to form the South Atlantic Ocean. As the breakup of Gondwanaland continued, Madagascar ended up as an island to the east of Africa. The discovery of

Where have the remains of a *Carcharodontosaurus* been found?

The *Carcharodontosaurus* (meaning "shark-tooth lizard") was a large theropod that rivaled the *Tyrannosaurus rex* in size. Fossils of this dinosaur were found in Morocco; the animal lived around 110 to 90 million years ago. The skull alone measured almost 5.5 feet (1.7 meters). The first *Carcharodontosaurus* was discovered in 1931, and was identified from an incomplete North African skull and a few bones. The remains were later destroyed during World War II. The most recent remains were found in 1996 and are even larger than the old specimen.

Majungatholus, and its similarity to dinosaurs in India and South America, made this scenario unlikely. How could this group of dinosaurs get from South America to Madagascar without first going through Africa?

To answer this question, scientists modified the sequence by which Gondwanaland broke up. Now they believe the landmass of Africa broke off from Gondwanaland first, becoming isolated from the rest of the supercontinent. South America, and the Indian subcontinent, which included Madagascar, remained connected to Antarctica as recently as 80 million years ago by means of land bridges. The group of dinosaurs to which *Majungatholus* belonged, as well as many other dinosaurs, could have migrated freely from South America to India and Madagascar by way of Antarctica. This would account for the remains found in those continents, and also for the lack of them in Africa.

FAMOUS PALEONTOLOGISTS OUTSIDE NORTH AMERICA

Who was **Gideon Mantell**?

Gideon Mantell (1784–1856), an English country doctor and fossil collector, was the first to recognize certain fossils as giant reptiles. The well known story (but perhaps not completely true) is that Mantell's wife, Mary Ann, found some fossilized teeth in rocks along the roadside while accompanying her husband on a house call. (Some people believe that Mantell actually found the fossils.) The rocks had come from the Bestede Quarry in Cuckfield, Sussex, England.

In 1822, Mantell's examination of these teeth, and subsequent remains from the same area, led him to the first reconstruction of what is now known as a dinosaur. In 1825, a year after William Buckland's published description of *Megalosaurus,* Mantell published his own description of this ancient reptile. He named it *Iguanodon,* or "iguana tooth," because the teeth, though much larger, matched those of this modern lizard. Mantell subsequently used a pictorial representation of the *Iguanodon* on the coat of arms for his residence, Maidstone, in Kent, England. The town of Maidstone also has the *Iguanodon* embedded on its coat of arms.

Mantell was responsible for other fossil discoveries. And he scientifically described the first known dinosaur skin in 1852. This was from the forelimb of a *Pelorosaurus becklesii*.

Who is **Rinchen Barsbold**?

Rinchen Barsbold is a Mongolian paleontologist who has named many recently discovered dinosaurs from China. He has named the *Adasaurus* (1983), *Ansermimus* (1988), *Conchoraptor* (1986), the family Enigmosauridae (1983), *Enigmosaurus* (1983), *Gallimimus* (1972), *Garudimimus* and the family Garudimimidae (1981), *Harpymimus* and the family Harpymimidae (1984), *Ingenia* (1981), the family Ingeniidae (1986), the family Oviraptoridae (1976), and the suborder Segnosauria (1980). The duck-billed dinosaur *Barsboldia* (1981) was named in honor of Barsbold.

Who is **Peter M. Galton**?

Peter M. Galton is a British paleontologist working in the United States; he is known for naming several dinosaurs, including the *Dracopelta* and *Bugenasaura,* and for naming the order Herrerasauria. With Robert Bakker, he championed the cladistic theory to show that birds are modern-day dinosaurs. Galton showed that the *Hypsilophodon* did not live in trees; that hadrosaurs did not drag their tails (the tail was used as a counterbalance for the head); and that some dinosaurs, such as the pachycephalosaurs, would butt their heads together like modern-day rams.

HOW TO FIND DINOSAUR BONES

What is the **first precaution** you should take before searching for **dinosaur fossils**?

The first and most important precaution to take before searching for dinosaur fossils (or any other fossils) is to make sure you have permission to search the area. In addition, verify that you can collect and keep any specimens you find on the land. Without permission, you may be arrested and charged with trespassing, destruction of property, or even theft. It is much easier to ask permission beforehand than to deal with legal proceedings, arrest, or jail-time afterwards. Always find out what person, organization, or government agency owns the property.

How do **paleontologists**, or even amateur collectors, **find dinosaur fossils**?

The key to finding dinosaur fossils (or fossils of any kind) is to search in the right types of rocks. For fossils, this generally means sedimentary rock, which are rocks composed of materials such as sand or mud that were deposited in a lake, river, or ocean. However, not all sedimentary rock contains fossils, much less dinosaur fossils. The right combination of factors must have existed for a dinosaur to have been transformed into a fossil in sedimentary rock.

To find dinosaur fossils, paleontologists must locate sedimentary rocks laid down during the right time. In the case of dinosaurs that means periods within the Mesozoic era. For example, if the paleontologist is interested in Jurassic period dinosaurs, then he or she should be looking for rocks deposited during that time, not the Cretaceous period. Once the specific age of the rocks is determined, the location of such rocks can be determined from a geologic map of the area. These

Amateur fossil hunters can obtain topographical and other detailed maps from the U.S. Geological Survey at little cost (iStock).

maps pinpoint the locations of the various rock types exposed at the surface, and shows an area's topography (height of the land).

Once a location with the right type of rock is determined, the paleontologist explores the area on foot, checking for exposed rock, and features that may have led to exposed rock, such as folds and faults. The paleontologist also looks for areas where erosion—due to action by water, wind, or even humans—continually wears away the sedimentary rock. This ensures a continuing exposure of the rock and any fossils within the rock.

The last step is to continually search the area, which often leads to spotting an exposed bone or other fossil part. This may all sound very simple, but it is not: patience and perseverance are the key ingredients at this stage of the search. Once the paleontologist makes a dinosaur or other fossil find, then the excavation, transportation, and restoration processes begin.

How can **geologic maps** be obtained?

Geologic maps can be obtained through many sources. In particular, there are topographic maps put out by the U.S. Geological Survey. These maps can also be obtained from the government agency's Internet site at http://www.usgs.gov/pubprod/. Also check the Web sites of the state you are interested in because many times the state's geological surveys will have maps available. There are also somewhat useable maps of certain areas of the country, including terrain, highway, and satellite maps, on the site "Google Maps" (currently at http://maps.google.com/maps).

How is **modern technology** helping **map** and **study** dinosaur fossil **finds**?

At a dig site, the exact location of fossil discoveries can now be determined using a Global Positioning System (GPS), an instrument that uses satellite technology to pinpoint a location on Earth. This instrument is now readily available to even amateur fossil hunters; for professionals (and sometimes advanced amateurs) GPS use eliminates errors due to poor maps, shifting landmarks, and inaccurate compass readings. To further understand uncovered fossils, the orientation and distribution of the bones in all three dimensions can be obtained using Electronic Distance Measurement (EDM) and other survey devices. (There are also some high-resolution GPS systems capable of providing this data.) These instruments minimize the errors inherent in using compasses and tape measures.

There is another advantage to using these techniques: the data can be fed directly into a computer. With the help of such programs as Geographic Information Systems (GIS) and Computer-Aided Design (CAD), a three-dimensional map of a quarry site, showing the location and orientation of all the bones, can be generated. The paleontologist can then study the site from different orientations, attempting to answer questions such as, "What social structure did these dinosaurs have?" Or, "What caused dinosaur bones to be concentrated in this location?"

Where are **dinosaur dig sites** usually **located**?

Dinosaur remains can be found worldwide—from the barren deserts of Mongolia to the cold slopes of the Antarctic mountains. This is because, at the time dinosaurs roamed the land, all of today's landmasses were connected or close by, allowing dinosaurs to freely move about.

Still, all dinosaur dig sites have something in common: The action of natural—or sometimes human—agents has eroded the land, exposing the buried fossil-bearing rock. In many cases, the best place to discover the first bones that signal a major find is where these erosion processes continue today. In the Gobi Desert, the passing of another sandstorm means a fresh batch of bones will be waiting on the surface. Bases of sea cliffs, where the water batters the rock during high tides or storms, will have new fossils exposed. Areas of heavy downpours, flash flooding, and excavation in a commercial quarry are all good places to find the bones that will trigger the start of the formal dig process.

How are **dinosaur fossils distinguishable** from other fossils?

There are a number of ways to find out if your fossil is from a dinosaur. If you have the desire, there are numerous fossil guides and books to help you determine the identity of your fossil. In this approach, you will have to acquire some knowledge of taxonomy (the classification of plants and animals), as well as a general knowledge of biology and geology. This is a good way to learn for yourself, but it can be time-consuming and overwhelming for the beginner.

Another way is to have a more experienced fossil collector help with the identification process. Try a local fossil collecting club (and you may want to join, too). Also, a local university or nearby natural history museum usually has someone who will help identify the fossil for you, although sometimes a fee may be charged for this service. If they can't identify the fossil, they may be able to suggest someone who can help.

If your fossil does not correspond to anything known, it might be a new dinosaur species. In this case, your finding is very important to scientific knowledge, and you may be asked to donate the specimen to the museum or university for their collection—not only for the collection but for additional scientific study. Your name might even be used as the basis for the scientific name of the new species. Also, you might be asked to assist with further excavation at the dig site.

What relatively new method is being used to determine a fossil bone's original burial place?

A relatively new geochemical technique is now being used to determine a fossil bone's original burial site. This technique analyzes the rare earth elements present in bones and the surrounding sedimentary rock (rare earth elements are normally present in rocks and soil in small amounts). Dinosaur bones contain calcium phosphate (apatite) and proteins. When the animal dies, the protein rots away. After burial, the calcium phosphate reforms into a slightly different crystalline structure, and rare earth elements in the surrounding soil may replace some of the calcium ions in the bone structure. The relative proportions of the different rare earth elements present in a bone are a signature that is established soon after burial; and they remain fixed. If the rare earth element signature in a bone matches the surrounding sediment, it is likely that this was the original site of burial. However, if the signature of the bone is different from the surrounding sediment, then it is likely the bone was transported from its original burial site.

What are some **examples** in which **amateurs** found dinosaur **sites**?

One good example is the site of what is now the Mygatt-Moore Quarry near Fruita, Colorado. It was discovered on a late March hike in 1981 by Grand Junction, Colorado, residents Pete and Marilyn Mygatt and J.D. and Vanetta Moore. They were amateur rock and fossil hunters who had "cabin fever" that day and decided to go for a hike near the Utah border. During a lunch break, Pete Mygatt noticed a rock and picked it up. It split apart, revealing a partial tail vertebra of what was later identified as an *Apatosaurus*. This site is now named the Mygatt-Moore Quarry, and has yielded eight species of dinosaurs, including *Mymoorapelta,* a small armored dinosaur, and the first ankylosaur from the Jurassic period found in North America.

Another example is Rob Gaston, a local Fruita, Colorado, artist who found some of the earliest dinosaur tracks in western Colorado. His discoveries led to the discovery of the Gaston Quarry, where the *Utahraptor* was subsequently found.

Still another find: Christopher Wolfe, an eight-year-old third grader from Phoenix, Arizona, discovered the remains of the oldest known horned dinosaur during a trip to western New Mexico. As he climbed up a hill, he was attracted to a blackish purple object on the ground, which turned out to be a fossilized piece of the small horn that protected the dinosaur's eye. The rest of the fossilized remains included jaw parts, teeth, brain case, and other pieces. The dinosaur was approximately 90 million years old—the oldest known horned dinosaur to that date—and was named *Zuniceratops christopheri* after its discoverer.

DIGGING UP FOSSILS

What is a **dig site**?

A dig site is a localized area where numerous fossil remains are found and excavated by paleontologists. For example, if a herd of dinosaurs drowned while crossing a flood-swollen river, their bodies could have been deposited in a bend of the river. From there, their bodies would have been covered with mud, and fossilization would have taken place. Millions of years later, if a fossil collector discovered a few exposed fossils—and subsequent exploratory digging uncovered a large amount of fossils—then the area would become an active dig site.

A site where excavation is currently ongoing (or was worked in the past) is generally referred to as a quarry; after all, you are digging into the rock similar to a rock quarry collecting rock for commercial purposes, such as limestone or marble. Many times these quarries are named after the collectors who first found the fossil remains, such as the Mygatt-Moore Dinosaur Quarry near the Colorado and Utah borders; others are named after nearby towns.

How do **paleontologists dig** for dinosaurs?

Once an initial bone find has been made, evaluated, and the decision made to dig further, the process of excavating the rest of the bones commences. This is, contrary to the perception given in the media, a long, hard, labor-intensive practice, especially if the dinosaur was large and a complete skeleton is present. The overlying rock must first be removed, using appropriate tools. These tools can range from dynamite, bulldozers, and jackhammers to picks and shovels. Once this overlying layer has been removed, finer tools, such as dental picks and toothbrushes, are used to expose the upper bone surfaces. To prevent these exposed bones from drying out, cracking, or oxidizing, they are stabilized by applying appropriate chemical hardeners.

The exposed bone surfaces are then completely mapped, and a plan for the excavation of the entire skeleton is made. The first step is to isolate each bone, or group of bones, by digging vertical trenches around each, leaving a substantial thickness of rock in place for protection. Any bones exposed on the sides by this trenching should be stabilized, and each bone must be numbered with permanent ink and recorded on the map records.

The exposed bones on the top and sides are covered with layers of damp newspapers, tissues, or paper towels; then, the top and sides are covered with a jacket of plaster-soaked burlap strips. When dry, this jacket locks the bones into the rock, preventing any cracking or damage. Next, the rock on the underside of this block is carefully removed, a little bit at a time. Any exposed bones are again stabilized, and the newly exposed areas are jacketed. Pieces of wood or metal prop up the jacketed block as the amount of rock on the underside is slowly reduced.

Once this rock is small enough, and all the rest of the block has been jacketed, the block can be turned over. But before that, labels are placed on the jacket with permanent ink, indicating the mapping number for each bone inside, an orientation arrow, the date, the site name and number, and any other information needed by the museum for restoration. After the block has been turned over, any remaining exposed area of rock is jacketed and the bone or group of bones is ready for transportation to the museum.

Who first developed a **method** to **protect excavated** dinosaur **bones**?

Not only did a plethora of dinosaur bones result from the rivalry between Edward Drinker Cope (1840–1897) and Othniel Charles Marsh (1831–1899), but new methods to protect the precious excavated dinosaur bones were also developed, as well. For example, Cope's crew preserved fossils using jackets of burlap bags dipped in a paste of overcooked rice. This pasty rice mix was slathered over the burlap covering the bone; as it dried, it hardened enough to allow the bones to be safely shipped back to the eagerly awaiting scientists.

What types of **tools** are used at **dinosaur digs**?

There are a large variety of tools used at dig sites. But the excavation of any one fossil is a unique process and may only require a few of them. Also, different stages of the evacuation process may require different tools. With that in mind, here is a general list of the major tools that might come in handy at dig sites. (Also note: Some dig sites may be miles from roads, and tools must be carried in, so light-weight and multi-usefulness are important considerations):

1. Shovel or spade: A lightweight model for digging out loose material.
2. Geologic hammer: These hammers have a square at one end and a chisel or pick at the other end. They are indispensable, general purpose tools.
3. Club hammer: This tool is used for hitting heavy chisels. Geologic hammers are often substituted for this tool.
4. Rock saw and stonemason's chisels: These are chisels with an assortment of blade widths. They are used for removing rock from around a fossil.
5. Trowel or old knife: Trowels or old knives are used to scrap away soft rock.
6. Brushes: All types of brushes, from toothbrushes to paintbrushes, are used on a dig. They are good for removing loose rock.
7. Strainer or sieve: Strainers or sieves are useful for separating small fossil pieces from loose rock or washing samples.

What kind of **clothing and equipment** should you take into the field?

The best clothing and equipment would be similar to those you take hiking or rock-hunting. Of course, the clothing and equipment must be suitable for the local environment. For example, clothing for a dinosaur dig in the Gobi Desert would be much different than for one in Antarctica. And equipment needed for a day dig at a local quarry would be different than the equipment needed for a months-long expedition in a remote part of the world.

The following is just a general guide to clothing and equipment. It is by no means complete. If you are inexperienced in hiking or working in the outdoors, seek more specific advice from those who are—such as geologists, backpackers, paleontologists, and mountain-climbers. There are numerous books on preparing yourself for the outdoors, and outdoor stores can be gold mines of advice. Also, many organizations sponsoring dinosaur digs have lists of required clothing and equipment to use as guidelines.

It is very important to dress appropriately and to take all the necessary gear with you when you are fossil hunting, especially when you are in a remote area (Big Stock Photo).

1. Appropriate clothing for weather and conditions: For example, long sleeve shirts, T-shirts, shorts, long pants, sweaters, jackets, underwear. Control your temperature through the shedding or adding of multiple layers of clothing. Take enough for the length of the expedition.
2. Backpack: Used to carry tools, food, water, and extra items.
3. Sturdy boots: For hiking to the dig site; and protection from falling rocks and hard surfaces.
4. Gloves: For hand protection during digs. Also to keep hands warm in colder climates.
5. Safety helmet: Protection from falling rocks if working in an area with cliffs, or collecting in a working quarry.
6. Rain gear: Protection from getting wet; can also be used to stay warm.
7. Flashlight: For illumination at night or for under dark overhangs.
8. Hat: Protection from sun or rain. Use when there is no danger from falling rocks.
9. Sunglasses/sunblock: Protection from the Sun's ultraviolet rays.
10. Goggles: Eye protection from flying rock chips.
11. Canteens: For carrying water in remote areas.
12. Camera and video recorder: Record excavation of fossils.

13. Compass and map: For finding directions in remote locations.
14. First-aid kit: In case of injury or other medical emergencies.
15. Tent, sleeping bag, cooking gear: For shelter, rest, and cooking food at remote site.

PUTTING DINOSAURS TOGETHER

What happens to **excavated dinosaur bones** after they **reach a museum**?

What happens to excavated dinosaur bones depends on whether or not the museum or institute plans to exhibit the find in the near future. If there are no immediate plans for the bones, they will be placed in safe storage until time and funds are available for their preparation. However, if the skeleton is to be put on display relatively quickly—as in the case of a new, spectacular species—then the bones will go through a fairly standard preparation process.

Simply put, the bones must be removed from the encasing rock during the preparation process. Next, any missing parts must be identified and a substitute found. Lastly, the bones are attached together and the entire skeleton is mounted for display. This process is tedious and time-consuming. For example, it took seven years of work until the *Apatosaurus* at the American Museum of Natural History in New York City was exhibited to the public.

How are **dinosaur bones prepared** for study and display?

The process of preparing dinosaur bones excavated from the field is generally done in a laboratory, where a wide variety of tools and chemicals are available. The first step in preparation is to remove the rock from around the bones, using hand tools, dental picks, needles, microscopes, small pneumatic tools or anything else that does the job. This technique is laborious and exacting, and can only be mastered through hours of hands-on experience. Once the bones are exposed, they must be repaired, if needed, and stabilized to prevent further degradation. There are a wide variety of glues and adhesives that serve this purpose. Weak or cracked bones may require the addition of structural supports, such as fiberglass or steel bands.

How are **missing** dinosaur **bones restored** to a skeleton?

When a dinosaur skeleton that is to be exhibited has missing bones (which is the case for the vast majority of fossil finds), these bones must be restored. This is accomplished in a variety of ways: Some missing bones can be replaced with fossil bones or casts from another individual of the same dinosaur species. Often times, two or more partial skeletons can be combined to produce one complete skeleton. If these methods can not be used, the missing bones can be sculpted from a variety of materials, such as wood, epoxy, or ceramic.

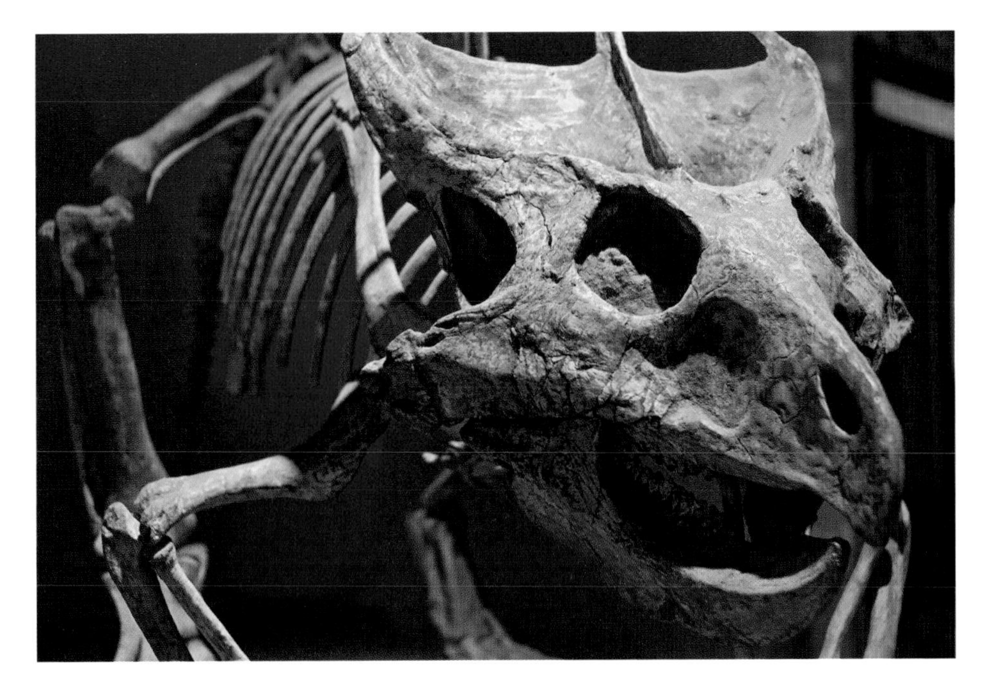

Mounting a dinosaur fossil for display is almost more art than science. Often missing bones need to be replaced with molds, and sometimes new discoveries make it necessary to rearrange the bones already mounted (Big Stock Photo).

How are **dinosaur bones mounted** for display?

Once the fossilized dinosaur bones have been prepared and stabilized—and any missing pieces obtained—the skeleton is ready to be free-mounted. The purpose of the mounted dinosaur is to display the skeleton as it might have looked in real life.

The first step, as in any large-scale project, is planning. Sketches and scale models are made, showing what the display will look like. Any variations to the posture—perhaps to reflect new information or to make the display more realistic—can be made at this stage, avoiding costly changes during actual assembly. The sketches and models will also show whether the skeleton will fit into its designated exhibition space. It would be very costly, time-consuming (not to mention embarrassing) to find out during assembly that the skeleton does not "fit." A good, final sketch can also be used as a guide during the actual assembly, as well as showing where extra support is needed.

Next, a strong steel armature (or armatures) is constructed and the individual dinosaur bones attached to it, in their proper places, of course. The armature is custom-made to provide enough support, but shaped to be unobtrusive. Because dinosaur bones are brittle, no stress can be placed on them; the armature is designed to support the weight of all the bones. Attachment of the bones to the armature is made using pins, bolts, or steel straps. Sometimes, it becomes necessary to hang cables from above to provide more support for parts of the skeleton.

Once the entire skeleton is mounted, there are still a few more details: the base on which the skeleton rests must be made visually appealing; barriers must be placed around the mount to protect it from the curious; and labels and display information created and positioned. At last, the dinosaur skeleton, which has remained hidden for millions of year, is ready to be viewed by the public.

Do **dinosaur bones** ever get **rearranged** after they are mounted?

Yes, dinosaur bones have been rearranged after they were mounted. This is because the field of dinosaur study is constantly changing as more bones are found, or more studies are completed. For example, in early 1999, scientists used a computer model of sauropod dinosaurs—dinosaurs with a neck up to 40 feet (12 meters) in length. Most of these animals were posed by museums with long S-shaped necks that would allow them to reach high into the tall trees to gather leaves. But the computer model indicated that the animals could not lift their heavy necks because the vertebrae were too heavy. These dinosaurs probably kept their necks straight out and may have chewed on lower-lying shrubs. Thus, many museums had to rearrange their sauropod mounts to fit the newest discovery.

How will **modern technology** help paleontologists **prepare, reproduce,** and **study** dinosaur fossils finds in the **near future**?

New technologies are rapidly changing the way paleontologists prepare, reproduce, and study dinosaur fossils. For example, Computed Tomography (CT) uses X-rays to generate a three-dimensional image of an internal structure of an object; it has already been used to determine if fossilized eggs contain baby dinosaur remains. Only those rocks containing fossils, or those eggs that contain baby dinosaurs, will be prepared, eliminating much of the destructive guess-work. In the near future, lab workers will also have available a three-dimensional image of a specimen to help them in the preparation process.

Once fossil remains have been prepared, precise measurements will be made using new instruments such as electronic calipers, or two- and three-dimensional digitizers. This data will be sent directly to a computer, which will guide machinery to automatically generate reproductions of the remains in materials such as metal or plastic. This process will make highly accurate casts available to more scientists (and the public) at lower costs. Another exciting possibility is that three-dimensional data might be obtained from such non-destructive techniques as CT, allowing paleontologists to make highly accurate reproductions of dinosaur fossils without ever removing the fragile bones from the encasing rock.

The research into dinosaur behavior and physiology will be greatly enhanced by the combined use of three-dimensional imaging, modeling, and virtual reality. Scientists will be able to study individual specimens, or even complete skeletons, from any angle or view, including from the inside looking out. And with data stored on computers, paleontologists will have much quicker access to rare specimens data. Such systems as the World Wide Web now allow scientists to study a rare specimen on the computer without having to travel to the few museums and institutions that house the actual fossils. Interestingly enough, even our computer games have been affected by such computer advances, with dinosaur games showing more "realistic" versions of these creatures.

LEARNING PALEONTOLOGY

Where can a person obtain more **education** in **dinosaur paleontology**?

If you desire more education and experience in dinosaur paleontology, there are numerous opportunities available at all levels. Your local college or university might offer courses in this specific subject, or the field of paleontology in general. Most colleges will allow a person to audit a course, or even take courses for credit. If you are really ambitious, you might want to take courses that could lead to a degree in paleontology, with an emphasis on dinosaurs. Some organizations offer formal instruction in conjunction with dinosaur digs, often for college credit.

Because of the interest in fossil collecting in general—and dinosaurs in particular—some museums and colleges are starting programs geared toward "professionalizing" the amateur fossil collector. These programs, which are intended to give the amateur the same level of practical field knowledge as the professional paleontologist, can be certification programs, or programs that lead to an associate's degree.

What **formal studies** would lead to my becoming a **professional paleontologist**?

The field of paleontology is an interdisciplinary field, requiring knowledge from many other fields of science, such as biology, geology, physics, and chemistry. Because the field requires a broad range of knowledge, there are relatively few educational institutions that offer degrees in paleontology, or in the more specialized field of dinosaur paleontology.

In high school, the best strategy would be to take as many science and math courses as possible, such as physics, chemistry, biology, geology, computers, and mathematics. A foreign language would also be helpful. The student should also read as much as possible on fossils and dinosaurs; visit museums with dinosaur displays; talk to other dinosaur paleontologists for information; and perhaps volunteer to participate in dinosaur digs.

In college, the more science courses you attend, the better. Most paleontologists have degrees in zoology or geology, and some have degrees in both. Zoology is useful for understanding the biology and taxonomy of animals, in this case dinosaurs; geology is needed to understand the local fossil environment, as well as to interpret the natural processes that occurred when the animal lived. A double major in zoology and geology would be ideal, but if that is not possible a major in one and minor

Check out Web sites like the Canadian Fossil Discovery Center to learn about how you might be able to participate in a dig.

in the other would suffice. Any other course to broaden the student's knowledge would be useful, such as statistical analysis and ecology. Conversing with professional paleontologists would give the student a working knowledge about the field of paleontology, and what is required to succeed in the field.

Most professional paleontologists have advanced degrees, and a master's or Ph.D. in paleontology can be obtained at a few colleges and universities. The student should carefully research the particular emphasis at their chosen college; the student should also check out the academic interests of the college faculty, making sure there is a match with the student's own interests.

What kind of work does a **professional paleontologist do**?

The work of a professional paleontologist varies, depending on where he or she works. Most paleontologists in the United States are college or university professors; some also work in museums, as independent consultants, or for government surveys. In general, professional paleontologists conduct research and write and publish academic research papers. They also curate, catalogue, and inventory fossils in a museum or university. They can run a research program, teach, or engage in a combination of these activities. Most of this work requires an advanced degree, although there are notable exceptions. For people with undergraduate degrees, the work can include preparing fossils, excavating and collecting fossils, mounting specimens for display, and casting specimens.

Where can I get **information** about **participating** in **dinosaur digs**?

So you want to get your hands dirty and actually dig for dinosaur fossils? As a novice, you will need education, training, and experience. The best way to obtain this is to participate in an organized dig. Depending on the program, there are opportunities ranging from one-day digs in pleasant surroundings to longer expeditions in places like Mongolia. Some programs include formal instruction for college credit, so check the details of the program in which you are interested.

The choices of dinosaur digs are limitless and can sometimes seem overwhelming. The following represents only a sampling of the programs offered by museums, education institutions, and private concerns, and their respective Web sites. Information about other opportunities can be found on the Internet, or in the classified ad sections of magazines dealing with paleontology, science, or nature. (Note: Some of the institutions listed below charge fees for participating in the dinosaur digs. In addition, some of the contact information may change over time.

The Canadian Fossil Discovery Centre
111-B Gilmour St.
Morden, Manitoba, Canada R6M 1N9
Phone: 204-822-3406
E-mail: info@discoverfossils.com
Web site: http://www.discoverfossils.com/

Dino Digs
P.O. Box 20000
Grand Junction, CO 81502-5020
Phone: 1-888-488-DINO ext. 212
E-mail: jcron@westcomuseum.org
Web site: http://www.wcmuseum.org/

Judith River Dinosaur Institute
Box 51177
Billings, MT 59105
Phone: 406-696-5842
E-mail: jrdi@bresnan.net
Web site: http://www.montanadinosaurdigs.com/

Timescale Adventures
P.O. Box 786
Bynum, MT 59419
Phone: 1-800-238-6873 or 1-406-469-2314
E-mail: info@timescale.org
Web site: http://www.timescale.org/main.html

The Wyoming Dinosaur Center
110 Carter Ranch Rd.
P.O. Box 868
Thermopolis, WY 82443

Phone: 307-864-2997 or 800-455-DINO
E-mail: wdinoc@wyodino.org
Web site: http://server1.wyodino.org/programs/

What are **certification programs** in the field of dinosaur **paleontology**?

In response to the growing number of fossil collectors interested in dinosaur paleontology and field work, a number of museums and institutions are running programs designed to turn amateurs into "para-paleontologists." Students are trained to collect and prepare dinosaur fossils like professionals, while remaining amateur collectors. With this training, some people may make the transition into a full-time paleontological career.

What are some **examples of certification programs** offered?

There are several certification programs. For example, the Denver Museum of Natural History in Colorado has a Paleontology Certification Program for adults (17 years of age and older). It is offered to people who want to learn more about paleontology and develop skills in the collection and preservation of fossils. Established in 1990, the core certification program includes a series of required courses. These courses provide an introduction to the history of life revealed through the fossil record, as well as knowledge of the theories and techniques in paleontology. After obtaining the basic certificate of competency, which requires completing the mandatory classes, passing a final exam, and approval by a museum committee, the student can continue studies in fossil preparation or field work. After hands-on laboratory courses and a final exam, the student obtains a lab specialization certificate; or, after additional related class work, six days of actual field work, and a final exam, the student can obtain a field specialization certificate. Some students may want to pursue both areas of interest. Classes are usually held in the evenings, with field trips on the weekends. For more information about this program, contact the museum (or check out their Web site at: http://www.dmns.org/main/en/General/Education/AdultProgram/Certifications/.

Another group that offers a certification program is the Utah Friends of Paleontology. This group offers a certification program for volunteers that prepares them to work with professional paleontologists and to teach others in the community about the field. You can find out more about this group by contacting:

State Paleontologist
Utah Geological Survey
P.O. Box 146100
Salt Lake City, UT 84114-6100
Phone: (801) 537-3311
E-mail: marthahayden@utah.gov
Web site: http://geology.utah.gov/esp/paleo/paleovol.htm

RESOURCES

How can I **learn more about dinosaurs**?

There are numerous resources available on dinosaurs. Your local library and bookstore probably have good selections of the latest dinosaur books. There are museums located around the world with permanent exhibits of dinosaur fossils, skeletons, and reconstructions, and most are extremely informative. Sometimes there are television specials, or movies dealing with dinosaurs; many of these can also be obtained for home viewing (often from your local library). In addition, there are traveling exhibits of animated dinosaurs; and there are parks and trails around the world where you can travel in the footsteps of early dinosaur explorers, and maybe even observe digs still in process. In addition, if you can connect to the Internet (either at home or in a local library), there are innumerable sites dealing with dinosaurs and related subjects.

Where can I find **books about dinosaurs** at my local library?

Your local library has a wealth of information about dinosaurs; you just have to know where to look. A computer search of the online catalog system is a good place to start. Look under the subject: dinosaurs. Most of the adult nonfiction books on dinosaurs will be in the science section, with Dewey Decimal numbers ranging from 567.91 to 568.19. Most libraries make finding such books easy with the help of computer search programs. Ask your reference librarian for directions to this section if you need help. And do not forget the juvenile section of the library, whether you are a juvenile or an adult. Many juvenile sections of libraries seem to have more books on dinosaurs, albeit on a less technical level. The same Dewey Decimal numbers apply here: with a prefix J (for juvenile), look in the range 567.91 to 568.19.

Where can I find **books about dinosaurs** at my local **bookstore**?

Books about dinosaurs are usually found in the science and/or nature section of your local bookstore. There may also be books of a less technical nature (and often more pictorial) in a separate children's section.

What are some **books** dealing with **general dinosaur information**?

The following is a list of just a few of our favorite books—many of them classics—that contain general information about dinosaurs:

Farlow, James O., and M.K. Brett-Surman. *The Complete Dinosaur.* Bloomington: Indiana University Press, 1997. A classic, comprehensive, easy-to-use reference with illustrations, chronology, and glossary. Also, a list of science fiction and fantasy books about dinosaurs.

Fraser, Nicholas. *Dawn of the Dinosaur: Life in the Triassic*. Indiana: Indiana University Press, 2006. A book about the time before dinosaurs ruled Earth.

Horner, John R., and Edwin Dobb. *Dinosaur Lives: Unearthing an Evolutionary Saga*. New York: HarperCollins, 1997. Celebrated paleontologist Horner recounts his discoveries of dinosaur eggs, babies, and nests in this classic text. It also examines the impact dinosaurs have had on our lives.

Lambert, David. *Dangerous Dinosaurs Q and A: Everything You Never Knew About the Dinosaurs*. New York: DK Publishing, 2008. Considered a young person's book, but very readable for general audiences—by a well-known author of numerous dinosaur books.

Manning, Phillip. *Grave Secrets of Dinosaurs: Soft Tissues and Hard Science*. New York: Random House, 2008. Manning presents the most astonishing dinosaur fossil excavations of the past 100 years.

Richardson, Hazel, and David Norman. *Smithsonian Handbook: Dinosaurs and Other Prehistoric Creatures*. New York: DK Publishing, Inc., 2003. Only one of many dinosaur books by world-renowned paleontologist David Norman. This book covers all the dinosaurs and other prehistoric creatures, complete with full-color illustrations.

Weishampel, David B., and Luther Young. *Dinosaurs of the East Coast*. Baltimore, Maryland: Johns Hopkins University Press, 1996. A classic survey of East Coast dinosaur findings and their importance in recreating the fossil records of dinosaurs in the region, with more than 130 illustrations.

What are some **books** explaining **dinosaurs evolution** and **extinction**?

The question of dinosaur evolution and extinction has been addressed in numerous books. The following are some examples the reader might want to check out:

Alvarez, Walter. *T. rex and the Crater of Doom*. New Jersey: Princeton University Press, 1997. This is the classic work by the man who began the dinosaur extinction controversy by suggesting a link between a certain type of rock and the extinction of the dinosaurs.

Bakker, R.T. *Dinosaur Heresies: New Theories Unlocking the Mystery of the Dinosaurs and Their Extinction*. Reprint. New York: Kensington Publishing, 1996. Another classic dinosaur book that dispels common misconceptions about dinosaurs, presenting new evidence that the creatures were warm-blooded, agile, and intelligent.

Larson, Pedro, and Kenneth Carpenter, eds. *Tyrannosaurus Rex, The Tyrant King*. Indiana: Indiana University Press, 2008. Covers the most famous of dinosaurs, with a CD included.

Long, John, and Peter Schouten. *Feathered Dinosaurs: The Origin of Birds*. Oxford University Press, 2008. All you've ever wanted to know about the controversial subject of bird origins.

Poinar, George, and Roberta Poinar. *What Bugged the Dinosaurs?: Insects, Disease, and Death in the Cretaceous*. New Jersey: Princeton University Press, 2008.

The authors offer evidence of how insects directly and indirectly contributed to the dinosaurs' demise.

Are there any **books** about **dinosaur expeditions** and **paleontologists**?

There are numerous books describing exciting dinosaur expeditions and the fascinating scientists in the field of paleontology. The following is a representative sampling of the books on these subjects:

Colbert, Edwin H. *The Great Dinosaur Hunters and Their Discoveries.* Reprint. New York: Dover Publications, 1984. Includes chapters on first discoveries, skeletons in the earth, two evolutionary streams, the oldest dinosaurs, Jurassic giants of the western world, Canadian dinosaurs, and Asiatic dinosaurs.

There are numerous resources available for the amateur paleontologist to learn about discovering his or her own fossils, as well as university and other programs that train people to assist on digs without needing an advanced degree (Big Stock Photo).

Doescher, Rex A., ed. *Directory of Paleontologists of the World,* 5th ed. Lawrence, Kansas: International Palaeontological Association, 1989. Lists more than 7,000 paleontologists by name, office address, area of specialization or interest, and affiliation.

Horner, John R. *Dinosaurs under the Big Sky*. Missoula, MT: Mountain Press Publishing, 2003. World-famous paleontologist John Horner's book about his knowledge of Montana's dinosaurs and geology.

Jacobs, Louis L. *Quest for the African Dinosaurs: Ancient Roots of the Modern World.* New York: Villard Books, 1993. After discovering a major fossil site in Malawi (Africa), Jacobs and his team went on to identify 13 kinds of vertebrate animals that "give a window into the world of this part of Africa one hundred million years ago."

Manning, Phillip. *Dinomummy: The Life, Death, and Discovery of Dakota, a Dinosaur from Hell Creek*. New York: Kingfisher, 2007. The story of the most amazing mummified dinosaur ever found—from the Hell Creek formation in North Dakota—aptly dubbed "Dakota."

Novacek, Michael. *Dinosaurs of the Flaming Cliffs.* Illustrated by Ed Heck. New York: Anchor Books/Doubleday, 1996. Chronicles the groundbreaking discoveries made by one of the largest dinosaur expeditions of the late twentieth century.

Psihoyos, Louie, and John Knoebber. *Hunting Dinosaurs.* New York: Random House, 1994. This book recounts the experiences of paleontologists who have scoured remote lands in search of dinosaur fossils, with full-color photos, charts, and maps.

Are there any **dinosaur books** that are suitable for **children** and **families**?

There are a large number of books that are suitable for the whole family and children of all ages. The following is a listing of some of these books:

Bergen, David. *Life-Size Dinosaurs*. New York: Sterling Publishing, 2004. If you want to see the size comparison of dinosaurs with what we know today, try checking out the illustrations in this middle-reader book.

Chaneski, John. *Dinosaur Word Search*. New York: Sterling Publishing, 2004. If you ever wondered how to pronounce some of those dinosaur names, this is the book for you.

Dixon, Dougal. *Dougal Dixon's Amazing Dinosaurs: More Feathers, More Claws. Big Horns, Wide Jaws!* Boyds Mills Press, 2007. One of many Dougal Dixon dinosaur books for kids—look for several others, too.

Johnson, Jay. *Dinosaurs*. Learning Horizons, 2005. A book about dinosaurs for ages five through eight.

Malan, John, and Steve Parker. *Encyclopedia of Dinosaurs*. New York: Barnes & Noble, 2003. A book that introduces the young reader to 250 different ancient species.

Norman, David. *Eyewitness Dinosaur.* New York: DK Publishing, 2008. One of many David Norman books—this one is heavily illustrated with a CD and wall chart.

Parker, Steve. *Dinosaurs and How They Lived.* New York: Barnes & Noble, 2004. A hundred facts with artwork about the dinosaurs—for young readers.

Stevenson, Jay, and George R. McGhee. *The Complete Idiot's Guide to Dinosaurs*. Alpha Books, 1998. Although somewhat dated, the book offers a guide to dinosaurs, including descriptions of more than 300 known species.

DINOSAUR WEB SITES

What are some **good Web sites** about **dinosaurs**?

Below are some helpful Internet resources to help you explore the world of dinosaurs and paleontology. Please note that Web sites may change.

American Dinosaur Fossils Exchange

Web Address: http://www.americandinosaurfossilsexchange.com/

If you have some disposable income, check this Web site. They specialize in dinosaur reproductions, replicas, and even real dinosaur fossils. Some of their items include skeletons, eggs, and teeth. There is also a page that shows how a replica dinosaur is made.

Carnegie's Dinosaurs

Web Address: http://www.carnegiemnh.org/carnegiesdinosaurs/index.html

See the dinosaurs exhibited at the Carnegie Museum of Natural History, learn about a new species called oviraptosaur, and watch bone preparation at the PaleoLab—all without leaving your computer.

CM Studio

Web Address: http://www.cmstudio.com/

Ever yearn to have a life-size dinosaur sculpture on your front lawn? If so, check CM Studio, which specializes in realistic dinosaur reproductions. And if life-size is too much, they also have available scale models, available in resin or bronze castings. Be the envy of your neighborhood—or just scare the postman.

DinoDatabase.com

Web Address: http://www.dinodatabase.com/dinowhre.asp

This Web site has an extensive listing of where to see dinosaurs. There are separate pages for the United States, Africa, Asia, Australia, Canada, Europe, Japan, Mexico, and South America. Within each page, there are addresses, phone numbers, and brief descriptions of the individual attractions; some even have links to their own Web sites.

"Dino" Don's Dinosaur World

Web Address: http://www.dinodon.com/dinosaurworld.htm

Fun site hosted by dinosaur expert "Dino" Don; includes a dinosaur dictionary, informative facts, trivia, and videos.

Dinosaur Farm

Web Address: http://www.dinofarm.com/index.html

Do you want to live the dinosaur lifestyle then this is the place to go. The list is endless: toys, books, lunchboxes, clocks, costumes, bedding, games, etc., all with a dinosaur motif. It's definitely for people who live, breathe, and sleep dinosaurs.

Dinosaur Guide

Web Address: http://dsc.discovery.com/guides/dinosaur/dinosaur.html

The Discovery Channel has produced numerous interesting shows about dinosaurs. Their Web site has three-dimensional views from these shows, an interactive tour, and animations.

There are many fascinating and informative Web sites available on the Internet, such as Paleo Adventure, a site that actually helps people arrange dig site tours, camping trips, and more.

Dinosaur Hall

Web Address: http://www.ansp.org/museum/dinohall/index.php

Philadelphia's Academy of Natural Sciences has a variety of dinosaur fossils in its Hall, including theropods, ceratopsians, and hadrosaurs; there is also the Big Dig, a replica of the New Mexico Badlands where dinosaur hunters can dig for fossils. Their Web site has information about all these attractions, as well as the fossil prep lab.

The Dinosauria

Web Address: http://www.ucmp.berkeley.edu/diapsids/dinosaur.html

Learn about the fossil record, life history and ecology, systematics, and morphology of dinosaurs from the University of California's Museum of Paleontology.

Dinosaur National Monument

Web Address: http://www.nps.gov/dino/

This area spanning Colorado and Utah is part of the National Park Service; there is information about what to expect when visiting on the Web site. If you cannot make the trip, there is a virtual tour of the Cliff Face of the Douglass Quarry, and a multi-media exhibit showing the inside of the Visitor's Center.

Dinosaurnews

Web Address: http://www.dinosaurnews.org/

A webzine since 1998, providing the latest dinosaur headlines from around the world.

Dinosaur Provincial Park

Web Address: http://www.tpr.alberta.ca/parks/dinosaur/event_desc.asp

Located in Alberta, Canada, Dinosaur Provincial Park is well worth a pilgrimage by dinosaur lovers. But until you can actually get there, their Web site will allow you to have a virtual visit, and to drool over scheduled activities such as the Fossil Safari Hike and the Centrosaurus Bone Bed Hike.

The Dinosaur Society

Web Address: http://www.dinosaursociety.com/

The Dinosaur Society bills itself as "dedicated to promoting an interest in everything dinosaur related." Its Web site reflects this, with pages devoted to dinosaur publications, news, finds, events and exhibits, and links.

Natural History Museum

Web Address: http://www.nhm.ac.uk/nature-online/life/dinosaurs-other-extinct-creatures/

London's Natural History Museum hosts a Web site with a section on dinosaurs containing, among other items, news articles, a dinosaur directory, and videos.

New Mexico Museum of Natural History and Science

Web Address: http://www.nmnaturalhistory.org/

New Mexico is a hotbed of dinosaur discovery, and the Museum reflects this activity. Permanent exhibits include Dawn of the Dinosaurs, Jurassic Super-Giants, and New Mexico's Seacoast. Check out the DynaTheater if they're playing "Dinosaurs Alive"—it's a treat for dinosaur lovers. The Web site also has an informative page about New Mexico's dinosaurs from the Triassic, Jurassic, and Cretaceous.

PaleoAdventures

Web Address: http://www.paleoadventures.com/index.html

This is the place when you want to get out from behind the computer and into the field. PaleoAdventures hosts dinosaur dig site tours, educational programs, and, for teachers, a field camp. They also sell genuine dinosaur teeth and bones that they've dug and prepared themselves.

Prehistoric Times

Web Address: http://www.prehistorictimes.com/

A full color quarterly magazine for the enthusiast and collector, Prehistoric Times has everything you need to be informed about dinosaurs. There are reviews of books and model kits, interviews with scientists and artists, the latest scientific discoveries, and gorgeous artwork.

Rocky Mountain Dinosaur Resource Center

Web Address: http://www.rmdrc.com/index.htm

The center has a great dinosaur display highlighting the late Cretaceous in North America. There is also a working fossil lab, theater, Children's Learning Center, and the Prehistoric Paradise Store.

Royal Tyrrell Museum

Web Address: http://www.tyrrellmuseum.com/

Located in Alberta, Canada, in the heart of the Badlands, this museum is another dinosaur lover's dream. Here, you can create your own cast of a fossil, spend a night with the dinosaurs, or hike the surrounding Badlands. There is a Dino Hall with almost 40 mounted skeletons; the Web site has a virtual tour.

ScienceDaily

Web Address: http://www.sciencedaily.com/search/?keyword=dinosaur

ScienceDaily bills itself as "your source for the latest research news." Searching on "dinosaur" will bring up a huge list of the latest dinosaur topics and stories. Bookmark this page and return often.

Sue at the Field Museum

Web Address: http://www.fieldmuseum.org/sue/index.html

Everything you ever wanted to know about the largest, best preserved, and most complete *T. rex* skeleton ever found.

The Wyoming Dinosaur Center

Web Address: http://www.wyodino.org/home/

A museum with over 20 dinosaurs displayed, dig site tours, gift shop, kids' digs, and opportunities to "Dig for a Day."

Zoom Dinosaurs

Web Address: http://www.enchantedlearning.com/subjects/dinosaurs/

Presented by Enchanted Learning, Zoom Dinosaurs is a comprehensive hypertext book about these animals for all ages and levels of knowledge. Topics include Anatomy and Behavior, Classification, and Extinction.

Index

Note: (ill.) indicates photos and illustrations.

C

E

F

G

H

I

J

K

L

M

N

O

S

U

V

W

Y–Z